Biotechnology in Clinical Medicine

First Edition

BY Beth Zielinski, PhD

cognella® | ACADEMIC PUBLISHING

Bassim Hamadeh, CEO and Publisher
Michael Simpson, Vice President of Acquisitions
Jamie Giganti, Senior Managing Editor
Jess Busch, Senior Graphic Designer
John Remington, Senior Field Acquisitions Editor
Monika Dziamka, Project Editor
Brian Fahey, Licensing Specialist
Rachel Singer, Associate Editor
Kat Ragudos, Interior Designer

Printed in the United States of America

ISBN: 978-1-63487-659-9 (pbk) / 978-1-63487-660-5 (br)

cognella® | ACADEMIC PUBLISHING

CONTENTS

CHAPTER 1
Introduction

a. What is Biotechnology?

A college student decides to buy some groceries, pick up an antibiotics prescription, and visit a grandparent who has just been implanted with a cardiac pacemaker. This student has just been exposed to at least three broad areas of biotechnology: genetic modification, pharmaceutical sciences, and substitutive medicine. Additional scientific disciplines based in the field of biotechnology include the biopharmaceutical sciences, stem cell therapies, cloning, and reproductive medicine. Although the term *biotechnology* has become fairly commonplace within the last decade, biotechnology has been in existence for thousands of years. So, what is biotechnology? It is the manipulation, manufacture, and use of biological materials (e.g., genomes, proteins, biological functions) for the purposes of developing products to be used in medical therapies and diagnostics, modified food products, bioremediation protocols, forensics, and bioterrorism. Even though some technologies—such as human identification through the use of fingerprints (forensic biotechnology) and the use of molds and herbal medicines to treat infections (early pharmaceuticals)—date back thousands of years, we will begin with a discussion of biotechnology in the nineteenth century.

b. The Age of Infection

Throughout time, humans have lived with the constant threat of microbial infection. Those individuals who did not succumb to primary infections developed immunities to them and passed these immunities on to their offspring. Once thought to be the result of malevolent spirits and

FIGURE 1.1: 18th Century Blood Letting
Michael Bernhard Valentini / Copyright in the Public Domain.

harmful entities located in the blood and other bodily fluids, patients with infections were often treated with a combination of spiritual exorcisms and bloodlettings (Figure 1.1). These remedies, in many cases, led to the deaths of patients and the perceptions that illnesses were causally related to good and evil, with evil being further linked to a lack of community status and economic prowess. Inquiry-based investigations into the biological origins of infections during the nineteenth and twentieth centuries led to significant breakthroughs in science that transformed medicine and public perceptions.

London's cholera epidemic during the late nineteenth century was significant to the development of modern scientific thought and ultimately modern biotechnology. Through a door-to-door census and methodical epidemiological mapping, Dr. John Snow was able to determine cholera's mode of transmission and London's index case for the 1854 cholera epidemic (1). The discovery that cholera is a waterborne illness that spreads easily in unsanitary conditions refuted the traditional miasma theory of disease. Disease, once thought to be associated with odors and an individual's socioeconomic status, could now be linked to a physical occurrence and location (Figure 1.2). As a result, sewer and other sanitation systems were constructed, which led to the design of the modern urban environment.

FIGURE 1.2: Epidemiological Map of London Cholera Epidemic, 1854
John Snow / Copyright in the Public Domain.

Approximately ten years following his epidemiological breakthrough, the germ theory of disease was established. Dr. Louis Pasteur showed that the growth of microorganisms in fermenting liquids, such as milk, was not due to spontaneous generation, but rather was the result of outside contamination. Although others had suggested this theory earlier, it was Pasteur who conducted the experiments and managed to persuade others that it was correct. He went on to invent a process

FIGURE 1.3: Carbolic steam spray used by Joseph Lister, England, 1866

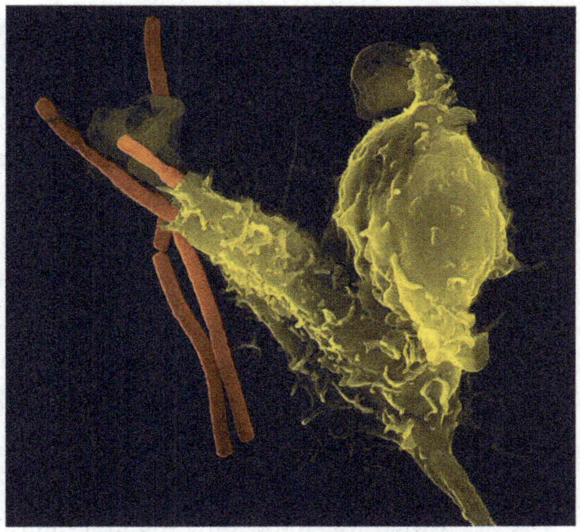

FIGURE 1.4: White Blood Cell Engulfing Anthrax Bacteria

in which liquids were heated to kill all bacteria. It is for this reason that we talk about milk having been "pasteurized" (2). Pasteur's work inspired Joseph Lister to develop antiseptic methods, the use of carbolic acid spray, in surgery just a few short years later in 1865 (Figure 1.3) (3). The field of microbiology had emerged, and with it the discovery that human diseases could be caused by bacterial pathogens.

In 1882, the scientist Robert Koch was the first to design a series of experiments to evaluate the germ theory of disease (4). In one series of experiments, Koch inoculated mice with *Bacillus anthracis* that he had extracted from infected sheep and then had grown in the laboratory. Inoculated mice became infected with the *Bacillus anthracis* and developed the disease anthrax, thus proving that disease is transmitted by microorganisms (Figure 1.4). Over the next ten years, several investigators, including Dr. Paul Ehrlich, identified a host of disease-causing organisms and demonstrated that antibodies were responsible in part for immunity to these organisms, thus paving the way for vaccine research. An early form of vaccination, variolation, had previously been developed in the early nineteenth century by Edward Jenner against the Variola (Figure 1.5). However, identification of smallpox did not occur for another twelve years, and development of the first commercialized vaccination against the polio virus was not to occur until the mid-twentieth century.

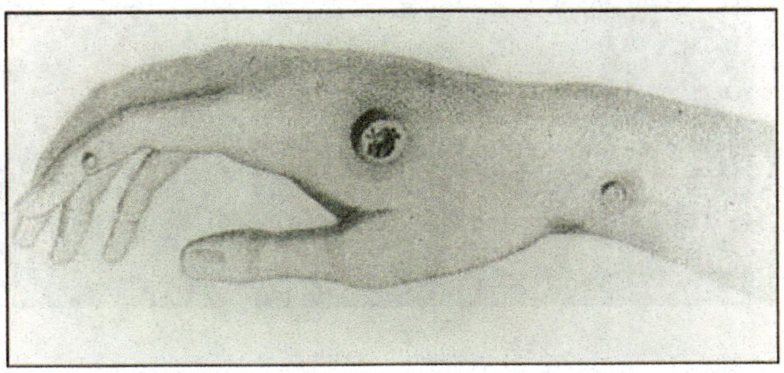

FIGURE 1.5: Variolation with Smallpox

c. Treating the Effects of War

Antibiotics

The significance of basic scientific research during the latter part of the nineteenth century and early part of the twentieth century cannot be overstated. Following the discoveries of the existence of pathogens and their relationships to human diseases, Sir Alexander Fleming, through a serendipitous finding, discovered the antibacterial properties of *Penicillium* mold (Figure 1.6) (5). This discovery eventually gave rise to the development of antibiotics as powerful tools against bacterial infections. Upon returning from a family vacation, Fleming observed that a mold, *Penicillium*, had grown on some of the petri dishes that had been inoculated with *Staphylococcus aureus* bacteria and had, in effect, killed the bacteria. These observations prompted him to write a scientific article on his discovery. Fleming named the mold's antibacterial agent *penicillin*, which was to become the first manufactured and commercialized antibiotic twelve years later. A commercialized version of penicillin became available during World War II through chemical modifications implemented by Howard Florey and German refugee Ernst Chain (Figure 1.7) (5). Many soldiers who suffered debilitating wounds during battle, which subsequently became

FIGURE 1.6: Sir Alexander Fleming at Work in his Microbiology Laboratory, 1943

UK Ministry of Information / Copyright in the Public Domain.

FIGURE 1.7: Marketing of Penicillin During World War II

Chemical Heritage Foundation / Copyright in the Public Domain.

infected and gangrenous, were saved from sepsis and imminent death by the administration of penicillin.

Prosthetics

Some of these soldiers returning from the war were left with upper- and lower-limb amputations and were unable to reacclimatize into society once they arrived home from the war. The return of these soldiers, just as soldiers returning from overseas deployment today, prompted investigations into ways of replacing lost limbs and optimizing prosthetics.

Prosthetics, used to replace missing upper and/or lower extremities, have been in existence for thousands of years. Ancient Egyptians used bone and wood as replacement materials for prosthetic digits (Figure 1.8). For at least a century, metal alloys have most commonly been used at the distal end of upper-limb prosthetics to allow for rudimentary grasping. During the mid-twentieth century, more functional devices were developed to replace these archaic designs. These prosthetic replacements, still in use today, rely on electrical signals generated in the remaining muscle groups of the stump (Figure 1.9). As these muscle groups contract, a generalized electrical field is generated, and electrical signals are transmitted through the skin to electrodes and then to wires within the prosthetic. These signals are ultimately transduced into mechanical movements at the distal portion of the prosthetic that allow the individual to grasp objects more readily. Although this is a vast improvement over poorly functional designs, these myoelectric devices are still heavy and cumbersome to attach and detach. Significant individualized training is required, and individuals fatigue from the required muscle contractions and the weight of the devices. Lower-limb prosthetics have followed a very similar path as upper-limb prosthetics with regard to patient practicality and function.

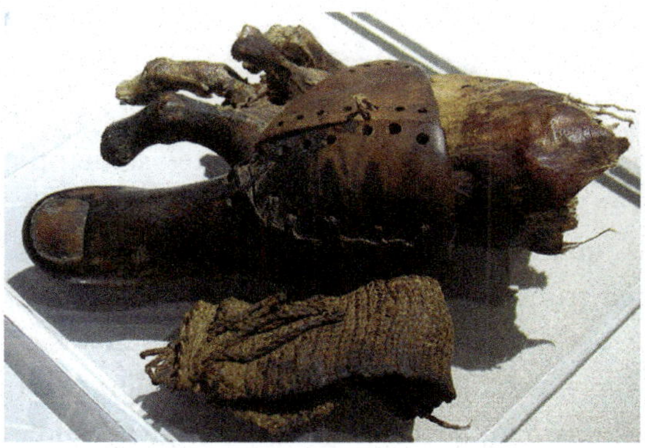

FIGURE 1.8: Ancient Egyptian Prosthetic Toe, 950-710 BC
Copyright © 2007 by Jon Bodsworth. Reprinted with permission

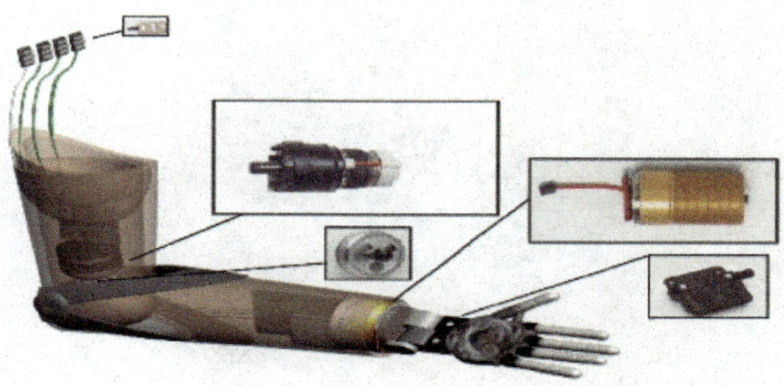

FIGURE 1.9: Myoelectric Upper Arm Prosthetic
Copyright © 2012 by Edgard Afonso Lamounier, Jr., Kenedy Lopes, Alexandre Cardoso, and Alcimar Barbosa Soares, (CC BY 4.0) at http://omicsonline.org/2155-9538/images/2155-9538-S1-010-g009.html.

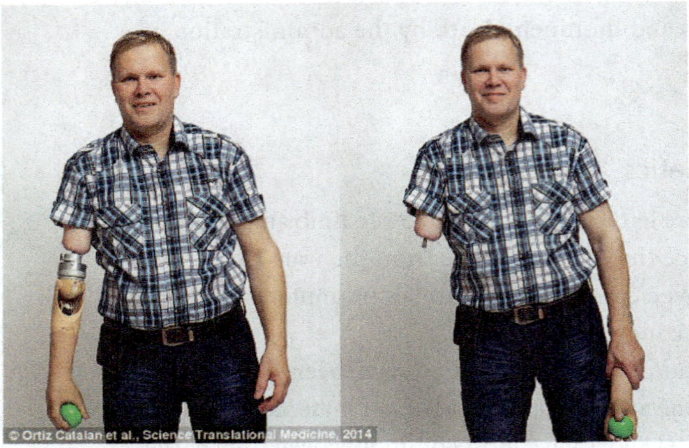

FIGURE 1.10: Recipient of Brain-Controlled Prosthetic Arm
Max Ortiz-Catalan, Bo Håkansson, and Rickard Brånemark, from Science Translational Medicine, vol. 6, no. 257. Copyright © 2014 by American Association for the Advancement of Science. Reprinted with permission.

The need for improved prosthetic designs was further demonstrated during the early 1960s. During this time, a significant number of infants who were exposed to the drug thalidomide during their first three months in utero were born with the congenital condition phocomelia, which results in malformation of the limbs (6). As these children grew old enough to be fitted with prosthetic limbs, biomechanical and compliance issues associated with the prosthetics became grossly apparent. The children refused to wear these devices because they were heavy, uncomfortable, and did not function well enough to provide a benefit.

The twenty-first century has heralded innovations in prosthetic design and practicality. The use of improvised explosive devices during the Iraq and Afghanistan wars has resulted in men and women returning to the United States with significant upper and lower limb amputations. This rise in the number of amputees has prompted increased funding for prosthetics research and development in both the public and private sectors. Devices are now lighter weight and agile, and within the last two years bionic limbs have been developed that contain neural interfaces that allow for the sense of touch (Figure 1.10) (7). Prosthetic innovation in the twenty-first century includes contributions not only from mechanical and electrical engineering, but also computer science, materials engineering, and the biological sciences.

FIGURE 1.11: Willow Tree Bark Herb
Copyright © by Depositphotos / Kalcutta.

d. Anxiety, Heart Disease, and the Age of Pharmaceuticals

Historical records left by Hippocrates sometime between 460 BC and 337 BC describe several medicinal therapies for pain relief, including extracts isolated from willow tree bark (Figure 1.11) (8). These records indicate some of the first attempts at isolating active ingredients from plant sources for the treatment of specific pathophysiological conditions. Since that time, humans have explored ways to cultivate medicinal remedies from botanical and animal sources. The modern pharmaceutical industry, however, did not come into existence until the late nineteenth and early twentieth centuries with the discoveries of penicillin and insulin. Methods of isolation, purification, and manufacture of therapeutic drugs were necessary in order to build companies that could provide drugs to the masses. During the late nineteenth century, Eli Lilly and Company was one of the first pharmaceutical companies founded. The early days of selling small quantities of elixirs and salves gave way to developing methods capable of manufacturing billions of doses of drugs, and in 1944, penicillin was mass produced for the soldiers fighting in World War II. Similarly, Bayer AG was also founded in the late nineteenth century and matured as the company began to produce and distribute the drug aspirin.

During the early twentieth century, research into new drug development skyrocketed, and by the time World War II had ended, efforts to create drugs to enhance health and lifestyle were underway. Patients now lived longer lives and the goals of new drug makers were to further increase life expectancies, treat and/or cure diseases, and improve overall wellness. These efforts included the development of drugs such as cholesterol medications to reduce the risk of heart disease as well as sedatives and stimulants to reduce life's daily stresses.

Although the pharmaceutical companies sought to design drugs to better the human condition, many drugs—such as thalidomide—were synthesized and distributed to patients without clear understandings of the potential conditions that could be treated. Clear knowledge of drug side effects was lacking at the time. In the early days of the pharmaceutical industry, pharmaceutical companies had little oversight, as the Food and Drug Administration (FDA) had just been founded in 1906. Many of the FDA's drug policies were not written and implemented until decades later as potentially life-threatening side effects from prescription drugs were uncovered.

As pharmaceutical companies matured, drug advertising, originally left to attending physicians, began to be targeted at consumers. Early print advertisements gave way to radio and television advertisements. The ads conveyed promises ranging from stress relief to cures for overactive and unruly children. Mothers in particular were targeted by ads because they were the individuals primarily responsible for child care and interactions with physicians and nurses. As the FDA's policies on drug development, testing, and distribution matured during the latter half of the twentieth century, so did the agency's oversight of industry advertising (Figure 1.12).

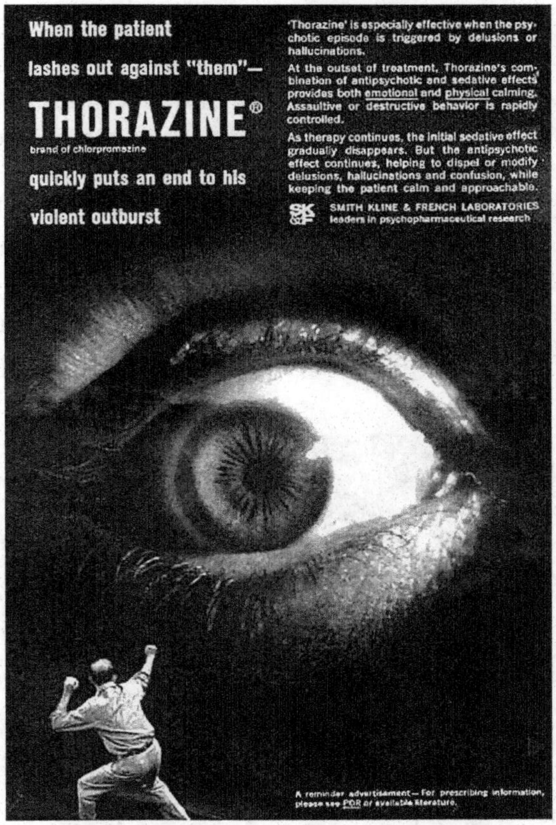

FIGURE 1.12: Early Pharmaceutical Advertising
Smith Kline & French Laboratories / Copyright in the Public Domain.

Today, the US Department of Justice and the FDA have put in place strict laws governing the use of advertising by the pharmaceutical industry (9). Steep fines and potential legal actions can now be imparted on those companies that do not abide by federal advertising laws.

In addition to increased government oversight, the twenty-first century has also heralded the arrival of more personalized strategies about patient care. Discoveries in the fields of genomics and proteomics have directed the generation-stratified pharmaceutical treatments and diagnostics, as well as engineered protein therapeutics and genetic manipulation. Additionally, the areas of organ transplantation, substitutive medicine, reproductive medicine, stem cell therapy, and regenerative medicine have all benefited from advances in the fields of genetics and DNA technologies.

e. Organ Transplantation and Substitutive Medicine

Organ Transplantation

In 2012, over one hundred thousand patients were on the waiting list for an organ donation (10). The number of patients waiting for an organ continues to increase, but the number of available viable organs is severely limited at approximately twenty-five hundred organs per year, excluding kidney donations, which include living donors (Figure 1.13). In part, limitations in available organs result from a lack of a sufficient number of donors and difficulties in matching donors with recipients. The waiting list for organ donations has increased every year since 1984 when the US Congress passed the National Organ Transplant Act in order to monitor the ethical use of organs and to address the country's organ shortage (11). Additionally, a centralized registry, operated by the United Network for Organ Sharing, was established to monitor patient organ matching and transplants, and the sale of human organs was outlawed (12). Establishment of a universal donor card system and required referral (which necessitates that hospital staff notify the local organ procurement organization of all patient deaths with subsequent follow-up by this organization with the potential donor's family) significantly standardized transplant protocols (13).

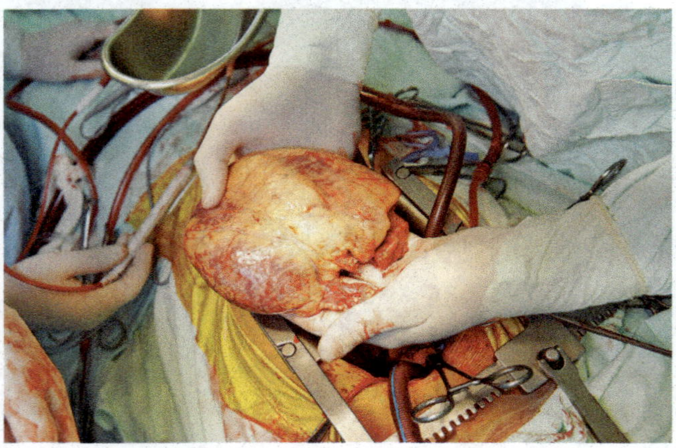

FIGURE 1.13: Cardiac Transplant

Copyright © by Depositphotos / kalinovsky.

Policies such as these would have been futile had it not been for the most significant achievement in transplant medicine: the development of immunosuppressant drugs to prevent chronic organ rejection (Table 1.1). In 1969, Jean-François Borel, a microbiologist working for Sandoz Laboratories in Switzerland, discovered cyclosporine, derived from a fungus found in soil in Norway, and in 1983, the FDA approved the use of this drug for use with organ

TABLE 1.1: Immunosuppressant Drugs

Drug	Mechanism of Action
Cyclosporine	Blocks the transcription of cytokine genes in activated T cells, thus preventing proliferation
Tacrolimus	Inhibits transcription of cytokine genes in activated T cells, thus preventing proliferation
Sirolimus	Blocks cell cycle progression at the juncture of G1 and S phases in T cells and B cells, thus inhibiting proliferation
Mycophenolate mofetil	Inhibits the purine synthesis pathway involved in the proliferation of T cells and B cells
Azathioprine	Blocks purine metabolism and DNA synthesis and prevents T cell and B cell proliferation

transplantation (14). Cyclosporine remains one of the most effective antirejection drugs available. The first successful kidney transplant was performed in 1954 by Drs. Joseph Murray and David Hume at Brigham Hospital in Boston, Massachusetts (15). The success of this transplantation was due to the fact that the donor and recipient were identical twins whose organs were complete biological matches. Techniques to improve organ viability through the development of temperature regulation and nutrient perfusion protocols have further advanced success in the field of organ transplantation. As a result, partial and full face transplants and complete limb transplantations have now been added to the list of more traditionally transplanted donor organs, such as kidneys, hearts, and lungs.

Substitutive Medicine

Although the process of replacing diseased organs with healthy donor organs is the gold standard in substitutive medicine, this standard cannot always be achieved. Replacement of physiological functions with synthetic or partially synthetic systems is a viable alternative. Substitutive systems contain synthetic and biological components that are capable of reestablishing the main function(s) of diseased organs and/or tissues (Figure 1.14). Biomaterial-based devices have been developed for treatments in cardiovascular, orthopedic, and reconstructive medicines. Materials such as Dacron polyester fiber and polyurethanes have been used in a variety of cardiovascular repair devices (16). Percutaneous interventions, for the treatment of both coronary and peripheral arterial blockages, employ a catheter system that deploys a polyurethane balloon to mechanically press the blockages against the walls of the arteries (Figure 1.15). Interventions such as these delay the need for more invasive measures, and for cardiac repair, specifically, delay the potential need for transplant. Other types of polymers, such as polyacrylonitrile, polysulfone, and regenerated cellulose, are used as dialyzing membranes for renal hemodialysis, allowing for extracorporeal renal support and increased life expectancy (17). Other medical conditions that

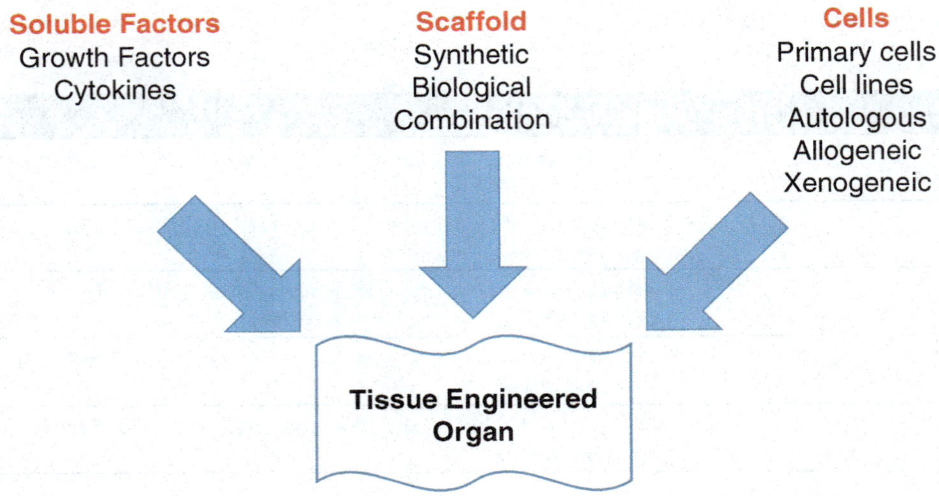

FIGURE 1.14: Essential Components for Tissue Engineering

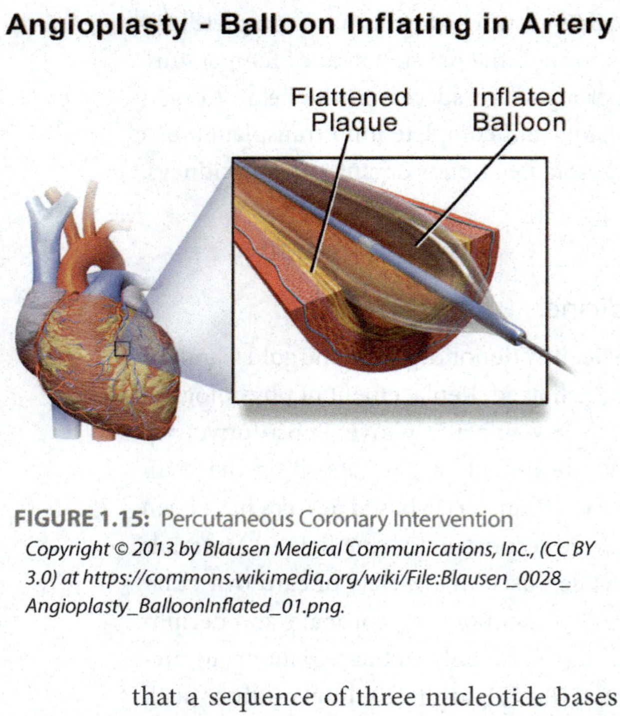

FIGURE 1.15: Percutaneous Coronary Intervention

are serious but not life-threatening and do not ultimately require organ transplantation can also be treated with devices manufactured from synthetic and partially synthetic materials. Prosthetic limbs and transdermal drug-delivery devices are two examples (Figures 1.16 and 1.17).

f. The Emergence of DNA-Based Medicine

The elucidation of DNA's structure by Watson, Crick, and Franklin during the early 1950s marked a turning point in biotechnology and medicine (18). Following on their discoveries, the steps involved in protein biosynthesis were delineated, and scientists demonstrated that a sequence of three nucleotide bases determined the structure of all twenty amino acids (Figure 1.18) (19). Less than a decade later, Paul Berg designed a technique to cut DNA strands and ligate these strands into a circular molecule of DNA (20). This demonstrated for the first time the capabilities of creating recombinant molecules, which soon led to the creation of recombinant organisms (Figure 1.19). Recombinant organisms are those organisms in which foreign

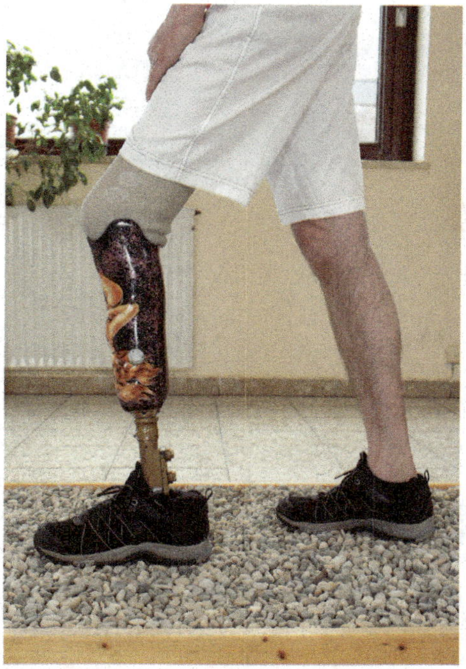

FIGURE 1.16: Lower Limb Prosthesis
Copyright © by Depositphotos / belahoche.

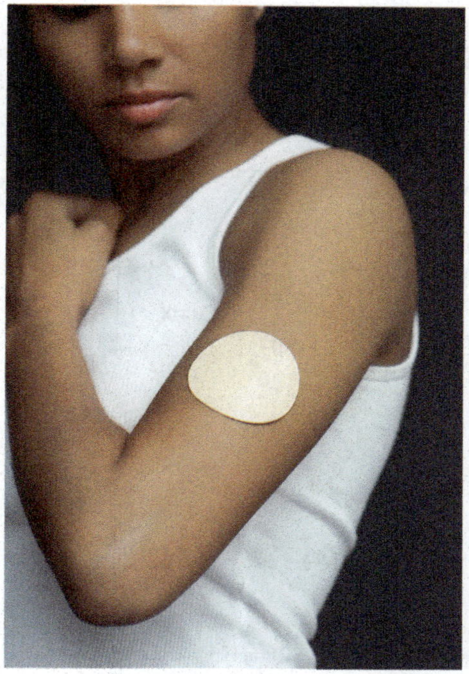

FIGURE 1.17: Transdermal Drug Delivery Patch
Copyright © by Depositphotos / londondeposit.

DNA has been inserted alongside native DNA, resulting in the organism's ability to transcribe and translate the newly introduced DNA into a protein. The first experiments demonstrating recombinant technology were in 1973 with the creation of recombinant bacteria containing both bacterial and viral genes with dual antibiotic resistance (21). Based upon this new technology, Genentech, Inc., a biotechnology company located in San Francisco, California, created a recombinant strain of bacteria carrying the gene for human somatostatin. This was the first synthetic recombinant gene system used to make a protein. This soon led to the development of recombinant insulin by Genentech, and hundreds of recombinant therapies for the treatments of diseases such as rheumatoid arthritis, hemophilia, and erythropoietin soon followed. Other fields that have benefited significantly from recombinant technologies are agricultural biotechnology, bioremediation, and the food sciences.

The idea that exogenous DNA could be inserted into bacterial and mammalian cells prompted investigators to transfer not only small pieces of DNA into individual cells but to transfer large pieces and whole constructs into entire eukaryotic organisms. Creating genetically altered organisms with the potential to produce sustained levels of foreign protein products—the science of transgenics—soon gave way to methods for transferring entire genomes from one organism to another. Transgenics led to cloning of complete organisms. The index case was the creation of a cloned ewe, Dolly, in 1997 (22). Dolly was created using a technique called somatic cell nuclear

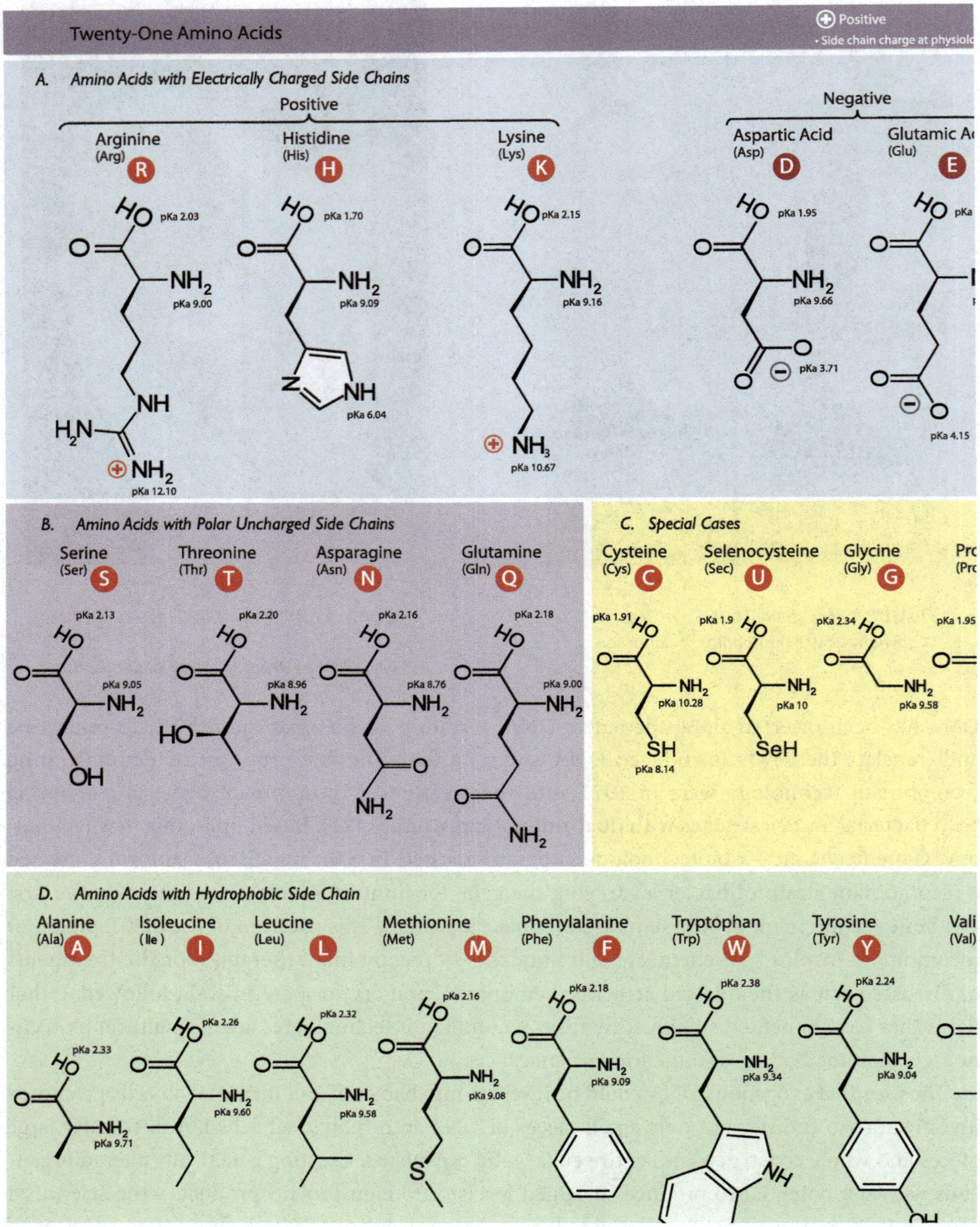

FIGURE 1.18: 20 Amino Acids

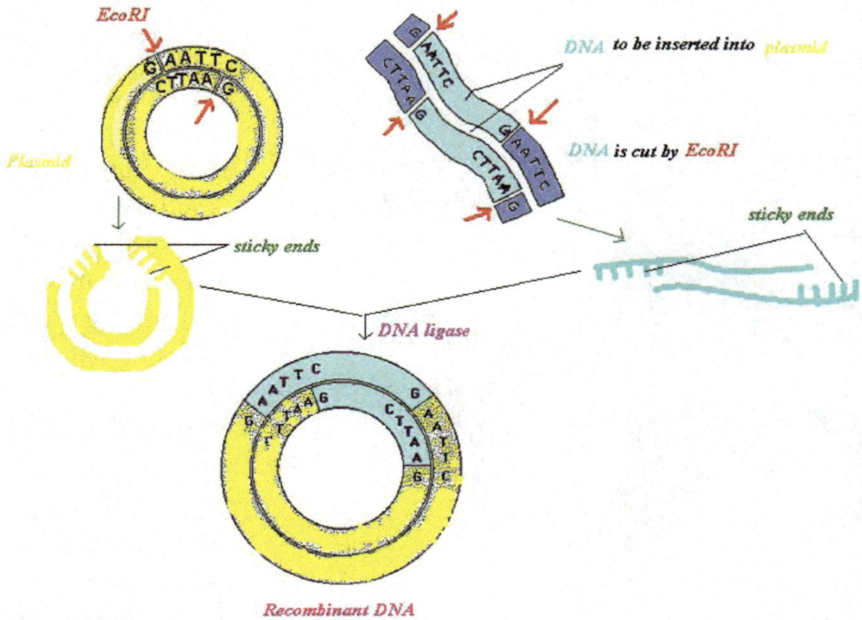

FIGURE 1.19: Recombinant DNA
Tinastella / Wikimedia Commons / Copyright in the Public Domain.

transfer that involves harvesting the entire nucleus from a somatic cell of one organism and inserting this nucleus, along with its entire genome, into the enucleated egg of another organism (Figure 1.20). Following electrical stimulation of the egg, cell division occurs and an embryo forms. The reconstructed embryo is then transferred into a surrogate for gestation and birth.

In a similar fashion, ex vivo human embryos can be created through in vitro fertilization. Human eggs, induced to mature in the female by exogenously administered hormones, can be removed, exposed to donor sperm in vitro, electrically stimulated to begin cell division, and then reimplanted into the female egg donor or a surrogate. The resultant fetus is considered a "test-tube baby." The first test-tube baby, Louise Brown, was born in Great Britain in 1978 (23). This event heralded the modern field of reproductive medicine. Since that time, technologies have developed that allow for the freezing and long-term storage of embryos, preimplantation genetic diagnostic testing of embryos, embryonic genetic

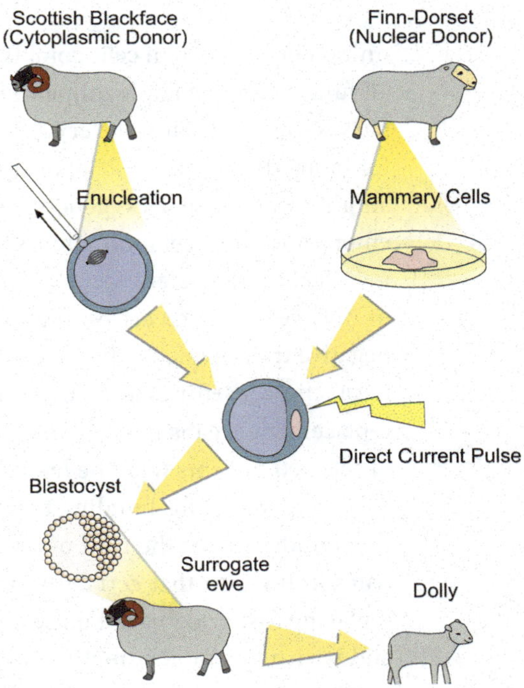

FIGURE 1.20: Cloning of Dolly
Squidonius / Wikimedia Commons / Copyright in the Public Domain.

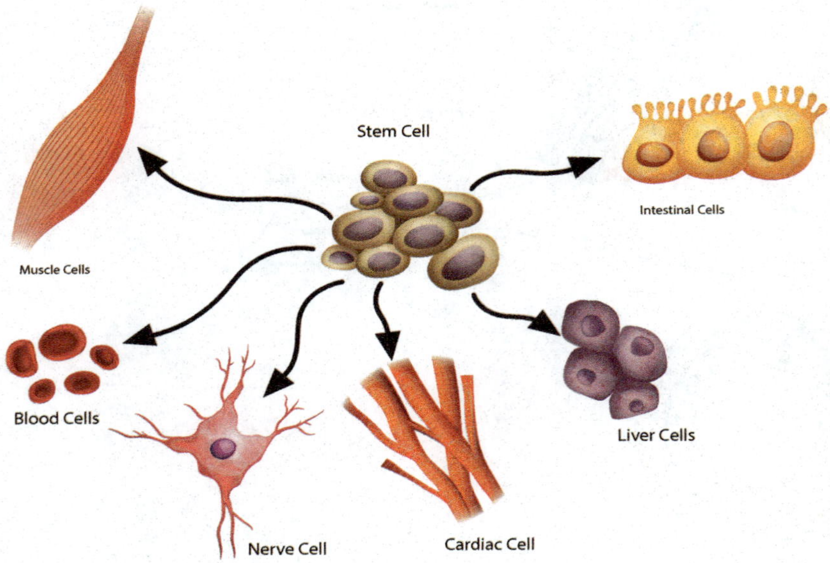

FIGURE 1.21: Potential of Embryonic Stem Cells
Copyright © by Depositphotos / blueringmedia.

manipulation, and in utero fetal surgery.

Cloning and specialized reproductive technologies have initiated a firestorm of controversy over the creation and manipulation of life, but they have also provided a base of knowledge for the disciplines of stem cell engineering and regenerative medicine. Stem cells, by definition, are cells that are capable of dividing to produce exact cellular copies, and when influenced by soluble factors and their external environments, are capable of producing a more differentiated progeny. Stem cells may be totipotent, pluripotent, or unipotent. Thus, stem cells combine a capacity for unlimited replication with a potential to inform all functions of an adult animal (Figure 1.21). Creation of a reconstructed clone such as Dolly and harvest of its cells during very early embryonic development is one way to create stem cell cultures for biomedical research. Harvesting cells from cultured biopsies obtained from adult tissue is yet another way, albeit a less genotypically powerful source (Figure 1.22). These two sources represent embryonic stem cell and adult stem cell populations, respectively. A third source of stem cells is induced pluripotent stem cells. These are adult cells that are induced, through the insertion of oncogenes, to revert back to a stem cell–like lineage. All three cell types have positive and negative traits, which will be discussed in more detail in a later chapter. Stem cells have been used in preclinical settings as drug delivery systems and as cellular building blocks for the creation of replacement organs in vitro. Recently, stem cells have been directly injected into damaged tissues of selected patients with the hopes of regenerating the tissues in situ (24).

The potential for building reparative and replacement tissues from this primary cell source is the foundation for the field of regenerative medicine. The theory of regenerative medicine originates in the idea that certain organisms, such as planarian flatworms and hydras, are capable of replacing lost or damaged body parts de novo. This potential lessens in more complex organisms and is virtually nonexistent in humans. By using primary cell sources, such as stem cells, soluble biological factors, and supportive scaffolds, investigators have shown that restorative tissues can be grown ex vivo and then implanted to replace functions of damaged tissues and organs (Figure 1.23) (25). A breadth of protocols exists that employs the triad of cells, biological factors,

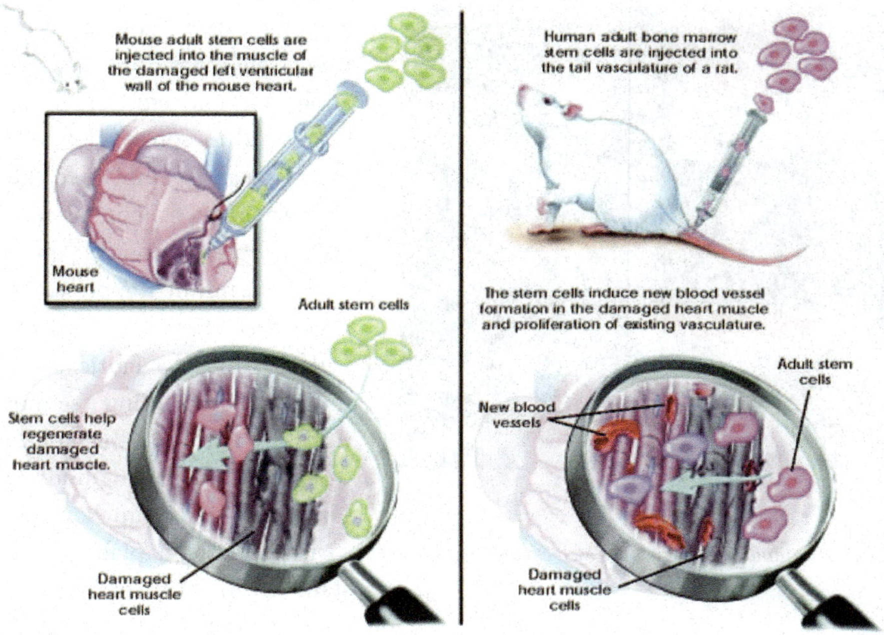

FIGURE 1.22: Adult Stem Cell Harvest and Application
NIH / Copyright in the Public Domain.

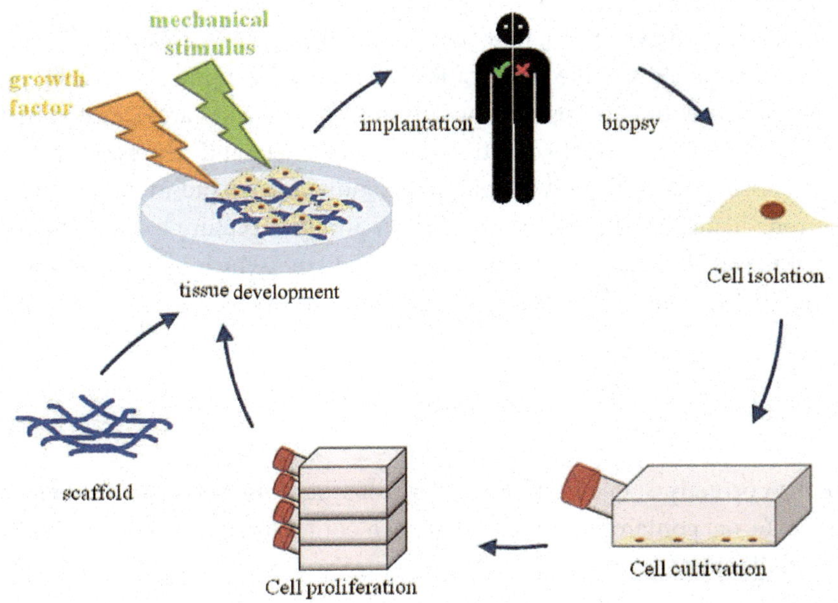

FIGURE 1.23: Tissue Engineering Using Stem Cells, Scaffolds and Growth Factors
Copyright © 2010 by HIA / Wikimedia Commons, (CC BY 3.0) at https://commons.wikimedia.org/wiki/File:Tissue_engineering_english.jpg.

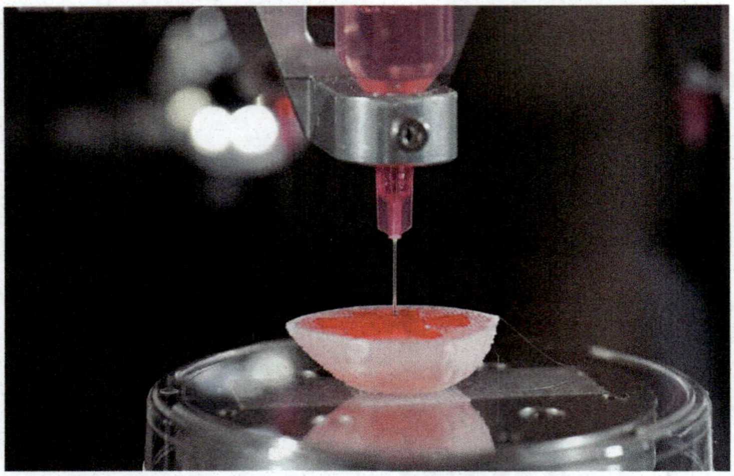

FIGURE 1.24: Bioprinting a Kidney

and scaffolds for ex vivo tissue construction. Cells harvested from autologous, allogeneic, and xenogeneic sources have been used to construct tissues that—in theory—can be implanted into humans as permanent functional replacements. Recent breakthroughs have yielded more streamlined approaches to actual tissue formation; the most recent innovation being the fabrication of tissues using inkjet printing technology (26). Anthony Atala and his colleagues have designed a novel method, *bioprinting*, for spraying mammalian cells from a cartridge onto a platform, much as a standard inkjet printer sprays ink onto paper (Figure 1.24) (27). Although the tissues at this early point do not contain all of the cellular layers necessary for full functionality, this proof of principle is the first step in attaining designer organs.

Biotechnology, a term that has grown in popularity over the last ten to fifteen years, is, in fact, a field that has existed, in modern form, for over a hundred years. The creation of biotechnology-based companies responsible for the purification of natural compounds and formulation of therapeutic drugs was truly the first significant step in global patient care. Development of more specialized drugs and devices to treat conditions such as anxiety and depression and to address the growing issues associated with limb amputation soon followed and increased the breadth of biotechnology in medicine. Finally, discovery of DNA and its structure led to the elucidation of the genetic code and establishment of genetics-based medicine, which has shaped the disciplines of reproductive medicine, stem cell technology, and regenerative medicine.

End-of-Chapter Questions with Answers

1. Which theory correctly states that the growth of microorganisms in fermenting liquids results from external contamination and that such contamination is responsible for human infections and diseases?

   Germ theory

2. What organism was the source of the first potent antimicrobial against *Staphylococcus aureus* bacteria?

   ***Penicillium* mold**

3. What two therapeutic drugs formed the basis of the modern-day pharmaceutical industry?
   Penicillin and insulin
4. Which federal agency regulates the testing, manufacture, and marketing of food, drugs, and medical devices within the United States?
   Food and Drug Administration (FDA)
5. Out of five thousand compounds tested in the early pre-discovery phase, how many of these qualify for preclinical testing?
   Two hundred fifty
6. What term refers to movement of a prosthetic device through the use of electrical signals generated by muscle contractions in the residual stump?
   Myoelectric
7. In the twenty-first century, what percentage of amputees in the United States are upper-limb amputees?
   10 percent
8. What is the average number of organs available each year for donation?
   Twenty-five hundred
9. What class of compounds revolutionized organ transplantation by suppressing chronic organ rejection reactions?
   Immunosuppressants
10. What is the name of percutaneous intervention for the treatment of coronary artery disease?
    Balloon angioplasty
11. What are the three critical components required for regenerative and substitutive medicine?
    Cells, scaffolds, and soluble factors
12. What was the first synthetic recombinant gene system used to make a therapeutic protein?
    Recombinant insulin
13. What type of cloning technique was used to create Dolly?
    Somatic cell nuclear transfer
14. What cellular characteristic defines the genotype and phenotype of embryonic stem cells?
    Unlimited capacity for undifferentiated cell division
15. The field of regenerative medicine is based on what theory?
    De novo regeneration of lost or damaged body parts by organisms such as planarians and hydras

Works Cited

1. Johnson, Steven. *The Ghost Map: The Story of London's Most Terrifying Epidemic—and How It Changed Science, Cities, and the Modern World.* New York: Riverhead, 2006.
2. Ullmann, Agnes. "Louis Pasteur (French Chemist and Microbiologist)." *Encyclopaedia Britannica Online.* Encyclopaedia Britannica.

3. Cartwright, Frederick F. "Joseph Lister, Baron Lister (British Surgeon)." *Encyclopaedia Britannica Online*. Encyclopaedia Britannica.
4. Waller, John. *The Discovery of the Germ: Twenty Years That Transformed the Way We Think about Disease*. New York: Columbia UP, 2002.
5. Kardos, Nelson, and Arnold L. Demain. "Penicillin: The Medicine with the Greatest Impact on Therapeutic Outcomes." *Applied Microbiology and Biotechnology* 92.4 (2011): 677–87.
6. Kim, J. H., and A. R. Scialli. "Thalidomide: The Tragedy of Birth Defects and the Effective Treatment of Disease." *Toxicological Sciences* 122.1 (2011): 1–6.
7. Judy, J. W. "Neural Interfaces for Upper-Limb Prosthesis Control: Opportunities to Improve Long-Term Reliability." *IEEE Pulse* 3.2 (2012): 57–60.
8. Benzie, Iris F. F., and Sissi Wachtel-Galor. "1." *Herbal Medicine: Biomolecular and Clinical Aspects*. Boca Raton, FL: CRC, 2011.
9. Greenslit, N. P., and T. J. Kaptchuk. "Antidepressants and Advertising: Psychopharmaceuticals in Crisis." *Yale Journal of Biology and Medicine* 85.1 (2012): 153–58.
10. "Uniting People and Information to Help Save Lives." *OPTN: Organ Procurement and Transplantation Network*.
11. "OPTN: Organ Procurement and Transplantation Network." *OPTN: Organ Procurement and Transplantation Network*.
12. "UNOS DonateLife." *UNOS*.
13. "Materials." *Organdonor.gov*.
14. Colombo, D., and E. Ammirati. "Cyclosporine in Transplantation—a History of Converging Timelines." *Journal of Biological Regulators and Homeostatic Agents* 25.4 (2011): 493–504.
15. Murray, Joseph E. "The First Successful Organ Transplants in Man." Nobel Lecture. 1990.
16. Chlupak, J., E. Filova, and L. Bacakova. "Blood Vessel Replacement: 50 Years of Development and Tissue Engineering Paradigms in Vascular Surgery." *Physiology Research* 58 (2009): S119–39. *Researchgate.net*.
17. Santoro, Antonio, and Gualtiero Guadagni. "Dialysis Membrane: From Convection to Adsorption." *Nephrology Dialysis Transplantation* 3.Suppl 1 (2010): I36–39.
18. Cobb, M. "1953: When Genes Became 'Information.'" *Cell* 153.3 (2013): 503–06.
19. Berg, Jeremy M. "50 Years of Protein Structure Determination." *50 Years of Protein Structure Determination*. NIH, Jan.-Feb. 2009.
20. Berg, Paul, and Janet E. Mertz. "Personal Reflections on the Origins and Emergence of Recombinant DNA Technology." *Genetics* 184.1 (2010): 9–17.
21. Cohen, S. N. "DNA Cloning: A Personal View after 40 Years." *Proceedings of the National Academy of Sciences USA* 110.39 (2013): 15521–5529.
22. Vos, Shantille. "Dolly, First Cloned Sheep." *Dolly, First Cloned Sheep*. Nova, Apr.-May 2004.
23. Henig, Robin Marantz. *Pandora's Baby: How the First Test Tube Babies Sparked the Reproductive Revolution*. Boston: Houghton Mifflin, 2004.
24. Fukuda, Keiichi, and Shinsuke Yuasa. *Cardiac Regeneration Using Stem Cells*. Boca Raton, FL: CRC Press, Taylor & Francis Group, 2013.

25. Murphy, S. V., and A. Atala. "Organ Engineering—Combining Stem Cells, Biomaterials, and Bioreactors to Produce Bioengineered Organs for Transplantation." *BioEssays; News and Reviews in Molecular, Cellular and Developmental Biology* 35.3 (2013): 163–72.
26. Ozbolat, I. T., and Yin Yu. "Bioprinting Toward Organ Fabrication: Challenges and Future Trends." *IEEE Transactions on Biomedical Engineering* 60.3 (2013): 691–99.
27. Binder, Kyle W., Arthur J. Allen, James J. Yoo, and Anthony Atala. "Drop-On-Demand Inkjet Bioprinting: A Primer." *Gene Therapy and Regulation* 06.01 (2011): 33.

CHAPTER 2
Pharmaceuticals

a. History of Drug Development

In 2011, the number of physician visits in the United States was approximately 1 billion, and from these visits patients were prescribed at least three prescriptions per visit (1). As of 2013, this translated into nearly 70 percent of the US population using prescription medication. These numbers are staggering and are indicative of the rising health care costs, not only for diagnosed illnesses, but also for adverse events associated with drug side effects. By 2020, health care costs will comprise 20 percent of the total gross domestic product (GDP) and will continue to rise due to an aging population and increasing dependence upon prescription medications (2).

FIGURE 2.1: Ephedra

Plant- and human-derived medicines have been used by civilizations for thousands of years. Chinese, Hindu, and Mediterranean civilizations were the first to extract medicinal preparations, including senna, castor oil, and ephedrine (3) (Figure 2.1). During this time of early drug discovery, the Greek physician Galen recognized the importance of drug purity, which has come to be the cornerstone of modern-day pharmaceutical science (4). Physicians in the sixteenth and seventeenth centuries created extractions of hundreds of new medicines, which led to the creation of the first comprehensive list of pharmaceutical preparations: the pharmacopoeia (5). The London Society of Apothecaries was soon formed in 1617 and provided the impetus for the emergence of independent pharmacies (6). In 1841, the Pharmaceutical Society of Great Britain was founded, thus providing a cohesive body to oversee the basic education and training of pharmacists (7). A precipitous rise in the isolation and purification of medicinal compounds occurred during the latter half of the nineteenth and early twentieth centuries. This paralleled development of the germ theory by Louis Pasteur, who confirmed that pathogens were responsible for disease states and the causative agents could potentially be eradicated using specific compounds (8). The perception of disease was forever changed by these discoveries. As techniques in general medicine and surgery developed, methods to optimize isolation and purification of pharmaceuticals became essential as compounds such as ether and phenol were used to anesthetize patients during medical procedures (9). Attempts to minimize toxic drug side effects associated with these and other compounds were made not only by drug isolation and purification methods, but also by recognizing the importance of drug dosage and delivery methods. By the late nineteenth century, many of the medicinal drugs were produced in pill and liquid formulations, which incentivized patient compliance. As the public demand for pharmaceuticals increased, pharmacists were unable to supply adequate yields, and drug companies were established. Pfizer, Inc. and Eli Lilly and Company were two of the earliest pharmaceutical companies to be established.

A child born in 1900 was expected to live to approximately forty-seven years of age. Over the past one hundred years, life expectancy in the United States has increased significantly, and as of 2013, life expectancy for women and men averaged eighty-one years and seventy-six years, respectively (10). These increases in life expectancies are due to lifestyle changes, and more importantly, by innovations in medicine and pharmaceuticals. During the late nineteenth and early twentieth centuries, as established drug companies came into existence, infections such as pneumonia and influenza were the leading causes of morbidity and mortality in the United States. Leading treatments at the time included mercury, digitalis, and quinine, all of which carried significant side effects and many times hastened, rather than retarded, disease progression and ultimate death. Correlations were made between inadequate compound purification and drug dosing, and these fledgling pharmaceutical companies began to assume the responsibilities of drug safety, efficacy, mass production, and marketing. Purified bovine insulin and several types of vaccines, including seasonal influenza vaccines, were some of the first mass-produced pharmaceuticals (Figure 2.2) (11). Penicillin, which was discovered by Alexander Fleming in 1928, was established as a powerful antibacterial by Howard Florey in 1939 and was manufactured for public use in 1943. Thus, the antibiotic penicillin became the first pharmaceutical blockbuster.

The next few decades realized the potential of pharmaceutical development, and indications such as anxiety, high blood pressure, and heart disease became targets for drug therapies.

Amphetamines, diuretics, and beta-blockers were formulated and manufactured for an American society that was becoming reliant on medications to treat chronic conditions. As life expectancies increased, so did the need for managed care, including recurrent physician visits and prescribed medications. Patients who would have succumbed to infections in the early twentieth century were effectively treated with antibiotics and now faced chronic diseases such as cardiovascular disease, neurodegenerative disorders, and cancer. Oral insulin formulations and recombinant human insulin replaced traditional bovine insulin, and early chemotherapeutic agents, such as mustard gas and methotrexate, were replaced by new antimitotic compounds and recombinant cytokine therapies (12). In the latter half of the twentieth century, novel pharmaceuticals continued to be developed, including modified versions of older drugs, extended-release formulations, and alternative delivery routes, such as transdermal patches. Now, in the twenty-first century, both industrial and academic scientists have introduced a personalized approach to pharmaceutical development that focuses on drug efficacy in subsets of patient populations; that is, their physiological and pathophysiological reactions to medications. These stratified and personalized models—which include the anticancer therapy Herceptin®, produced by Genentech, Inc.—are currently supplemental accepted standards of care (Figure 2.3) (13). Pharmaceutical science continues to strive toward biological- and genetic-based treatments that will eventually become the patient standard of care.

FIGURE 2.2: Large Scale Manufacture of Vaccines
Copyright © 1964 by Wellcome Images, (CC BY 4.0) at https://commons.wikimedia.org/wiki/File:Large_scale_manufacture_of_vaccine_Wellcome_L0046118.jpg.

b. Drug Sources

Pharmaceuticals can be divided into four main categories: terrestrial, marine, synthetic, and semisynthetic (Table 2.1). Compounds extracted from plant and marine sources have been used

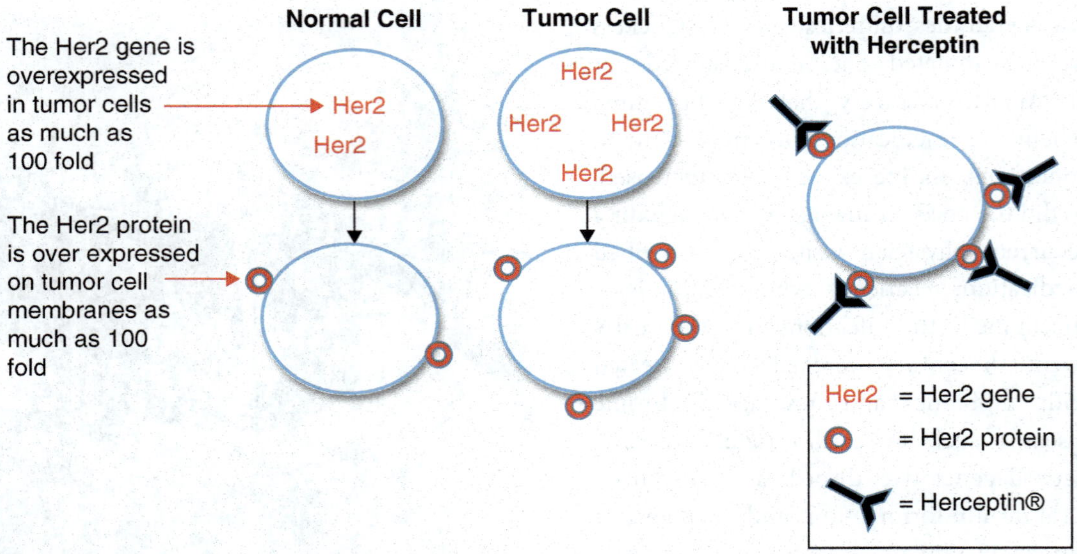

FIGURE 2.3: Herceptin®

for medicinal purposes for thousands of years. Digoxin, used to regulate heart function, and paclitaxel, a potent anticancer drug, are two examples of terrestrial-based drugs (Figures 2.4 and 2.5) (14). Pharmaceuticals can also be derived from natural templates and then modified in the laboratory in order to affect function and potency. Cocaine, a synthetically modified derivative from a compound found in the coca plant, was widely used in the late nineteenth century for maladies such as inflammation and broken bones. Amphetamine, a popular drug prescribed during the

TABLE 2.1: Pharmaceutical Categories Based on Source (original table)

Terrestrial: Drugs extracted from plant sources. *Digitalis is extracted from the foxglove plant.*	**Synthetic: Drugs chemically formulated in the laboratory.** *Thalidomide is chemically formulated and synthesized in a laboratory.*
Marine: Drugs extracted from marine organisms. *Cytarabine is derived from cells of Cryptotheca sp., a sponge.*	**Semisynthetic: Drugs extracted from natural sources and then chemically modified in the laboratory.** *Amphetamines are chemically modified from ephedra.*

FIGURE 2.4: The Foxglove Plant and Digoxin
Copyright © 2012 by Koosg / Wikimedia Commons, (CC BY-SA 3.0) at https://commons.wikimedia.org/wiki/File:Vingerh.jpg. Copyright © by Wellcome Images, (CC BY 4.0) at https://commons.wikimedia.org/wiki/File:Bottle_of_digoxin_tablets,_%27Tabloid%27_brand,_London,_England,_Wellcome_L0058212.jpg.

FIGURE 2.5: The Pacific Yew Tree is the Source of the Drug Paclitaxol
Copyright © by Depositphotos / Studiobarcelona. Paclitaxol: Calvero / Wikimedia Commons / Calvero / Wikimedia Commons / Copyright in the Public Domain.

1950s and 1960s, is also a semisynthetic drug that is a modified version of the naturally existing compound ephedra (15). Finally, purely synthetic pharmaceuticals are synthesized in the laboratory. Examples of these include tranquilizers and the controversial drug thalidomide (16). Thalidomide, originally synthesized by chemists working for the German pharmaceutical company Grünenthal, was prescribed by physicians to patients, including pregnant women, during the late 1950s and received over-the-counter (OTC) status in Europe in 1960 (Figure 2.6). A subset of women who had been prescribed thalidomide during their first trimester of pregnancy gave birth to babies who presented with the condition known as phocomelia. Phocomelia results from thalidomide's interference with vascular maturation during fetal development. Children born with this condition lack mature limbs and often are born completely missing one or more limbs. As approval for thalidomide was being requested in the United States, the FDA insisted on further safety trials and eventually denied approval (16). Thalidomide, however, is now being investigated as a potential anticancer drug due to its negative effects on developing vasculature. The repurposing of thalidomide presents an example of an additional source of drugs to treat disease.

FIGURE 2.6: Dr. Francis Kelsey Prohibits the Distribution of Thalidomide in the US in 1960
Frances O. Kelsey / Copyright in the Public Domain.
Klaus Hoffmeier / Copyright in the Public Domain.

c. The Drug Pipeline and the FDA

History

How are novel pharmaceuticals approved in the United States, and how long is the process? In order to fully answer these two questions, a brief history of the governmental body that regulates the pharmaceutical industry, the FDA, is necessary (17). As pharmaceutical companies began producing their first products during the early twentieth century, President Theodore Roosevelt signed into law the Food and Drug Act of 1906 (17). As a result, interstate transport of food that was adulterated with chemical additives or other contaminants and drugs that were misbranded as to their ingredients or indications was prohibited. This act was the first step in the establishment of the FDA. In 1938, President Franklin Delano Roosevelt signed into law the Food, Drug, and Cosmetic Act (FD&C Act), which further enforced governmental regulation of food and drugs (17). For the next three decades, drugs were reviewed for efficacy, and potential compounds could be withdrawn from the market as a result of drug-associated adverse events. In 1962, the Kefauver-Harris Amendment to the FD&C Act established further regulation, and pharmaceutical companies were required to show substantial efficacy as well as safety in three phases of clinical trials (18). This heralded the modern era of the FDA. The FDA's primary goal, concerning pharmaceuticals and medical devices, is to ensure patient safety and therapeutic benefit. FDA policies are continually being revised, and amendments such as the Hatch-Waxman Act of 1984 have included the creation of policies governing brand versus generic pharmaceuticals in terms of patent protection and bioequivalence (19). Post-market surveillance of approved pharmaceuticals by the FDA has also become critical as drugs move into the public market.

Drug Pipeline

The process of drug development begins, however, not with the FDA, but in academic and industry-based research laboratories. The process can be divided into five distinct stages: (1) prediscovery; (2) discovery and preclinical; (3) investigational new drug application (IND); (4) clinical trials; and (5) new drug application (NDA) and approval (Figure 2.7) (20). These stages are followed by post-market surveillance by the FDA (20).

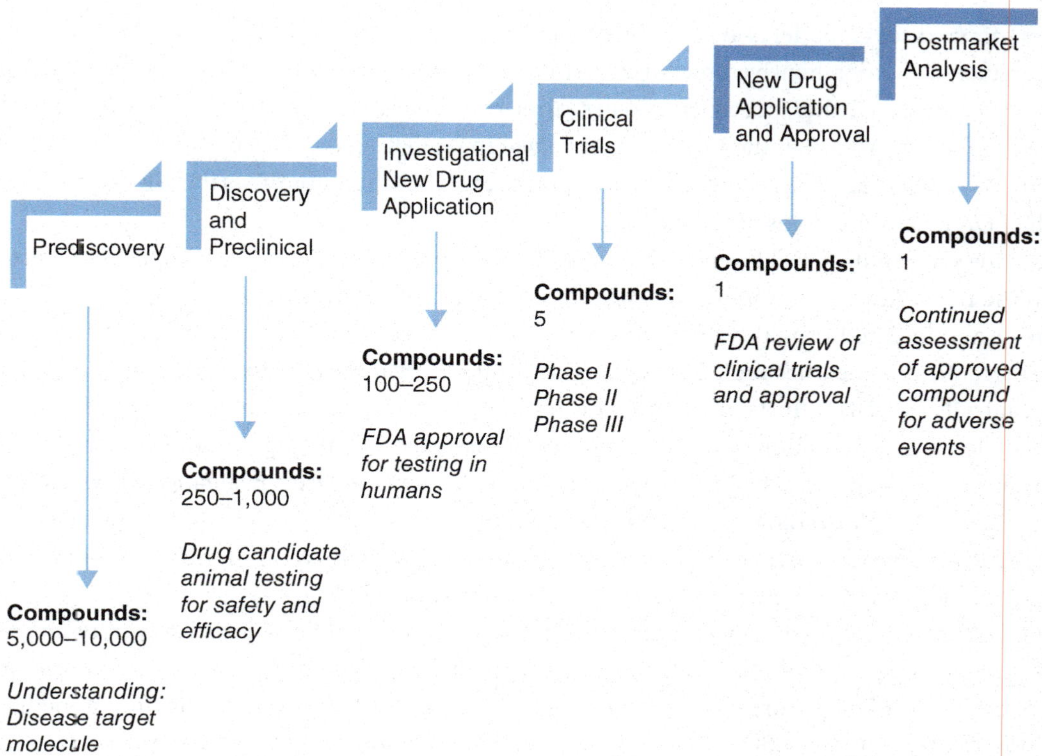

FIGURE 2.7: Drug Development Process (original schematic)

Drug discovery begins on the bench top in either academic or industry-based laboratories. Academic and industry-based laboratories follow parallel pathways to drug approval, albeit on different scales, with industry being on a much larger scale. The initial step in the process requires the principal investigator to gain knowledge and understanding of the disease state and target systems for which a therapeutic compound will be developed. In an industry-based setting, as many as five thousand compounds/ligands may be preliminarily screened for a specific disease state. Rapid screening includes analysis of binding affinities between these ligands and their protein targets and further in vitro testing with mammalian cells (Figure 2.8) (21). Compound structures, binding affinities, and initial cellular toxicity assays result in either inclusion or exclusion from the next step in the discovery phase. This next step, preclinical small animal testing, is focused on organismal data that shows safety and efficacy in small animal models (22). If a suitable

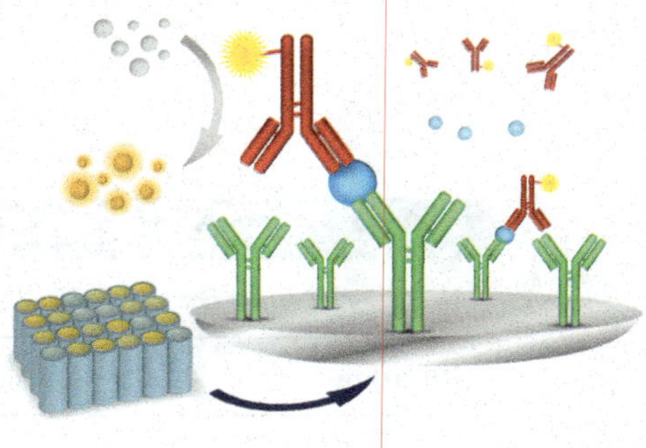

FIGURE 2.8: ELISA (Enzyme Linked Immunosorbent Assay) Used for Drug Screening

large-animal disease model is available, studies conducted in large-animal preclinical studies will be conducted. Comprehensive preclinical testing can take several years or more and reduces the number of potential therapeutic candidates from five thousand to approximately two hundred fifty compounds. Although the FDA is not involved in regulating these early studies, experiments should be conducted under good laboratory practices (GLP), because this data will be submitted to the FDA at a future point in the process.

At this stage in the process, approval to test the therapeutic compound in humans is requested from the FDA. Academic laboratories typically lack the funding necessary to complete studies in humans and look to hospital systems and pharmaceutical companies for collaboration. Likewise, pharmaceutical companies look to hospitals or contract organizations to collaborate on human clinical trials (23, 24). Efforts are focused on testing a single compound in human clinical trials. The application for initiation of human clinical trials is called an Investigational New Drug (IND) application and requires that FDA inspectors in the Center for Drug Evaluation and Research (CDER) or Center for Biologics Evaluation and Research (CBER) review preclinical data (25, 26). Investigators typically meet with FDA officials prior to IND submission in order to confirm the quantity and types of data that the FDA will eventually want to review.

Once IND approval has been granted, clinical trials commence with phase I safety trials and patient selection begins (Table 2.2). Patient recruitment can be difficult at all phases of study. Patients are often being asked to forego a treatment with documented safety and efficacy for something unknown. Clinicians are often aligned with one school of thought or one set of pharmaceutical companies and are hesitant to work in other venues. Additionally, with an increase in phase IV, post-market, trials and indications with limited populations, many of the available patients have already been signed up for other trials.

Human clinical trials begin with phase I trials in which healthy volunteers are exposed to small doses of the drug and monitored for acute (immediate) adverse effects. Once the clinical dosage is reached, pharmacokinetic adsorption, distribution, metabolism, and excretion (ADME) studies are conducted (Table 2.3). Efficacy is not an outcome parameter. Numbers are small—ten to twenty-five patients—and the trials take about six months to conduct and analyze. Patients, rather than healthy

TABLE 2.2: Clinical Trials (original table)

	Phase I	Phase II	Phase III
Number of subjects	10–25	100–500	1,000–10,000
Duration	< 6 months	6–12 months	3–6 years
Purpose	Acute safety	Dose response	Efficacy
Format	Healthy volunteers	Open label	Double blind, placebo control
Cost range	\$5–\$10 million	\$70–\$100 million	\$200–\$700 million

TABLE 2.3: Pharmacokinetic Testing during Drug Development: Adsorption, Distribution, Metabolism, Excretion (original table)

Adsorption	The amount of drug that passes into the bloodstream or target tissues by various routes
Distribution	The amount of drug transferred between tissue compartments in the body
Metabolism	The amount of drug broken down into metabolites within tissues in the body
Excretion	Amount of drug removed from the body either by the kidneys or in feces

volunteers, are tested when circumstances dictate (e.g., for chemotherapeutic agents, chronic pain). In this case, efficacy may be deduced, but this data is not considered convincing without a control.

Phase II trials are generally intended to evaluate the effect of the trial on the target disorder in humans (23–26). Participant numbers are small, one hundred to five hundred patients, but the individuals generally have the disease for which the drug is intended. Increasingly, trials are placebo controlled and can be extended up to twelve months. Dosage and schedule (e.g., once per day, twice per day, etc.) are optimized, and the therapeutic window is determined.

Phase III trials typically involve a main arm with approximately three thousand to five thousand patients to determine efficacy and then a collection of smaller trials to evaluate subsets or special concerns (23–26). Trials are prospective, randomized, placebo controlled, and double blinded (Table 2.4). *Prospective* means that the trial is designed in advance with agreed-upon protocols, study numbers, and outcome parameters. *Randomized* and *controlled* pertain to the patients who are assigned to a study group or control group by some random algorithm in order to avoid selection bias. In a double-blinded study, neither the patient nor the evaluator knows whether the patient is in the study arm or the control arm. This avoids *investigator bias*. Finally, most drugs are placebo controlled. Placebos are sugar or starch pills with the same appearance as the tested drugs. Patients receiving placebos often have objective response rates of 20 percent to 30 percent and sometimes higher. During the trial, an Independent Safety Monitoring Committee (ISMC) has access to unblinded data and is empowered to halt the trial if safety concerns arise. Typically, trials can extend from three to six years (25–26).

TABLE 2.4: Phase III Clinical Trial Parameters (original table)

Prospective	Therapeutic outcomes, as well as adverse events, are measured over the length of the study
Randomized	Study subjects are randomly assigned to receive either the study treatment or a placebo
Placebo-controlled	A cohort of subjects within the study receive a substance that has no therapeutic value in order to differentiate therapeutic efficacy of the bioactive drug therapy
Double blind	Researches and subjects do not know which treatment the subjects receive

Once clinical trials have been completed, investigators compile the data and summarize the results in a New Drug Application (NDA). Both data and comprehensive data reports are sent to the FDA, at which time reviewers determine whether the therapeutic should be approved or whether further testing is necessary. Currently, this may take up to two years due to the employee resources required by the FDA to perform this task. This is a fairly expedited process as compared to investigational drug reviews conducted by the FDA prior to 1992. In 1992, the Prescription Drug User Fee Act (PDUFA) was passed by Congress, which allows the FDA to collect fees from drug manufacturers to fund the new drug approval process (29). The PDUFA provides that the FDA be entitled to collect substantial application fees from drug manufacturers at the time an NDA is submitted, provided those funds are designated for use only in the CDER or CBER drug-approval activities. In order to continue collecting such fees, the FDA is required to meet certain performance benchmarks, primarily related to the speed of certain activities within the NDA review process. Fees can range from $600,000 to $1 million, and review times range from six to ten months. This entire drug pipeline process, beginning with drug development and preclinical trials and ending with FDA-approved human clinical trials and review, is very expensive, although it has proven to be fairly effective (30). The cost of clinical trials increases significantly with each phase. Phase I, II, and III trials cost $5–$10 million, $70–$100 million, and $200–$700 million, respectively (27, 28).

One main concern that continues to linger is the adverse events profile of the newly approved drugs once they are on the market and prescribed to large numbers of patients. Many dangerous and potentially life-threatening drug side effects cannot be determined in small clinical trial populations. In order to address such issues, both patients and physicians can report drug-related adverse events to the FDA because the FDA conducts post-market surveillance, which is considered a phase IV clinical trial. Additionally, drugs are only permitted to be prescribed for tested and approved indications and may only be marketed as such. Off-label use—prescription and use for indications other than those approved—does occur, however, and is the responsibility of the prescribing physician.

Drug Schedules

As prescriptions for stimulant and sedative drugs became more commonplace in the 1950s and 1960s, the potential for abuse increased. In order to regulate drug manufacturing, use, and distribution, the 91st US Congress passed the Controlled Substances Act (CSA) as Title II of the Comprehensive Drug Abuse Prevention and Control Act of 1970 (31). The act also served as the national implementing legislation for the Single Convention on Narcotic Drugs. With the passing of this act, a tiered system that outlines abuse potentials of certain drug categories was implemented. The tiers are defined as Schedules, and the Schedules range from I to V (Table 2.5) (32). Schedule I drugs such as opioid analgesics are those drugs that have high potential for abuse. These compounds have no currently accepted medical uses in treatments in the United States, and there is a lack of accepted safety data for uses under medical supervision.

TABLE 2.5: Drug Schedules (original table)

	Schedule I	Schedule II	Schedule III	Schedule IV	Schedule V
Description	No currently accepted medical use and high potential for abuse	Restricted medical use with high potential for abuse	Medical use with moderate potential for abuse	Medical use with low potential for abuse	Medical use with lowest potential for abuse
Examples	Heroin, lysergic acid diethylamide (LSD), marijuana (cannabis), 3,4-methylene-dioxymeth-amphetamine (Molly)	Hydromorphone (Dilaudid), meperidine (Demerol), oxycodone (OxyContin), Dexedrine, Adderall, and Ritalin	Combination products with less than 15 mg of hydrocodone per dosage unit (Vicodin), anabolic steroids	Xanax, Ambien, Valium	Cough preparations with less than 200 mg of codeine or per 100 mL (Robitussin), Lyrica

Schedule II drugs like oxycodone and many of the stimulant drugs, also have high potential for abuse. These drugs have currently accepted medical uses in treatments in the United States, albeit with severe restrictions. Abuses may lead to severe psychological or physical dependence. Schedules III through V drugs have lower potentials for abuse than Schedules I and II and are commonly prescribed. These include specific categories of stimulants, depressants, and opiates. Although potential for abuse is low, as stated, physical and/or psychological dependence can occur. The CSA is the federal US drug policy under which the manufacture, importation, possession, use, and distribution of certain substances is currently regulated.

d. Pharmaceutical Advertising

Marketing of newly approved drugs prior to the 1950s was not regulated by the FDA, and advertising was primarily controlled by the pharmaceutical companies that manufactured the specific compounds. As stimulants and psychiatric drugs became more prevalent in American society in the late 1950s, concerns were raised by then senator Estes Kefauver over misleading and insufficient information provided to consumers by the pharmaceutical companies. Consumers gained access to this information through literature distributed to physicians' offices and in medical journals. Although no laws were passed to address Senator Kefauver's concerns, the United Nations, in 1971, did implement a prohibition on psychotropic drug advertising (33). This did little, however, to curb pharmaceutical advertising.

The industry's abilities to advertise increased, and in the late 1980s, marketing of drugs was allowed in newspapers and magazines. This was the first step in direct-to-consumer (DTC)

TABLE 2.6: Pharmaceutical Advertisements (original table)

Product claim ad	Names the drug, the condition it treats, and talks about both its benefits and risks	*"If you suffer from disease X, drug Y may relieve your symptoms . . . drug Y should only be used by certain patients . . ."*
Reminder ad	Gives the drugs name but not its use	*"Ask your doctor . . ."*
Help-seeking ad	Describes the disease or condition but does not recommend or suggest specific drugs	*"You may be suffering from . . ."*

advertising (34, 35, 36). By 1997, broadcast advertising was approved, and in 1999, pharmaceutical companies began advertising stimulant drugs for children. The time of vetting the patient-disease-pharmaceutical relationship primarily through physicians came to an end, and potential patients now had opportunities to attempt self-diagnosis and treatment based on drug advertising. Advertisements aimed at describing disease states by using lay terminology armed individuals with basic medical information that warranted further investigations into marketed treatments.

In 2009, pharmaceutical companies spent a vast $4.8 billion in marketing to reach out to US consumers (37). Three types of DTC—product claim, help seeking, and reminder ads—became staples of pharmaceutical marketing (38) (Table 2.6). A *product claim ad* names a drug, the condition it treats, and talks about both its benefits and risks. A *help-seeking ad* describes a disease or condition but does not recommend or suggest specific drugs. A *reminder ad* gives the drug's name but not the drug's use. All three are used in print as well as broadcast advertising campaigns and have specific regulations that are followed, as mandated by the FDA. Print advertisements must contain all of the risk information described in the drug's FDA-approved label. Broadcast ads must directly state only the major risk information and are required to direct viewers to other sources of information.

Targeted advertising, "detailing," by pharmaceutical representatives directly to hospitals and physicians' offices is an additional marketing tool for pharmaceutical companies. This strategy is effective as well as expensive for the industry, with a price tag of approximately $4.8 billion in 2000 and increases in expenses estimated in the billions of dollars each year (39). In 2009, guidelines were put in place by the Pharmaceutical Research and Manufacturers of America (PhRMA) that prohibited the distribution of large and small gifts to physicians by pharmaceutical company representatives (40). Guidelines continue to be modified, and representatives are currently expected to be proficiently trained in pharmaceutical laws and regulations as well as industry codes of practice and ethics.

Although the FDA does not have authority to evaluate advertisements prior to public presentation, it does have the authority to evaluate existing ads and enforce restrictions if needed. Again, the industry must only market drugs for the conditions for which they are tested and approved by the FDA. Additionally, print and broadcast advertising campaigns must follow information guidelines. Penalties for not complying with FDA requirements can be categorized into three levels (41). The first order of reprimand is an *untitled letter* requesting that the company voluntarily withdraw the advertisement. If a company ignores the initial reprimand or a series of violations has occurred, then the company receives a *warning letter*. Additional violations or a complete lack of responsibility on the part of the pharmaceutical company eventually leads to investigation by the US Department of Justice and criminal prosecution. "In the largest settlement involving a pharmaceutical company, the British drug maker GlaxoSmithKline agreed to plead guilty to criminal charges and pay $3 billion in fines for promoting its best-selling antidepressants for unapproved uses and failing to report safety data about a top diabetes drug, federal prosecutors announced Monday. The agreement also includes civil penalties for improper marketing of a half-dozen other drugs" (42).

e. Antibiotics: The Cure and the Resistance

Penicillin was the first blockbuster drug to be developed, manufactured, and prescribed in the United States. *Staphylococcus aureus*, a bacterial strain that frequently colonized the flesh wounds of soldiers during World War II, was the first bacterial strain to be treated successfully with penicillin during the 1940s (43). As with the manufacture of penicillin, most antibiotics are produced by bacteria or fungi or are chemically modified derivatives of these natural antibiotics. Mechanisms of action vary among different antibiotic classes. The penicillins, cephalosporins, and vancomycins inhibit bacterial cell wall synthesis; the tetracyclines, erythromycins, and streptomycins destroy

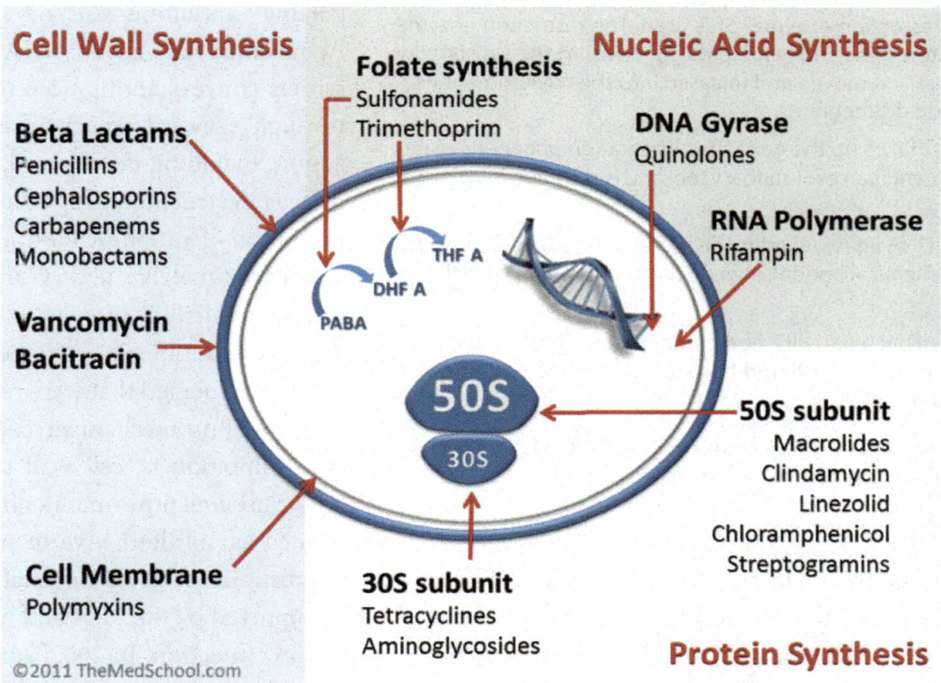

FIGURE 2.9: Common Antibiotics and Their Mechanisms of Action
Copyright © 2011 by Kendrick Johnson, (CC BY-SA 3.0) at http://commons.wikimedia.org/wiki/File:Antibiotics_Mechanisms_of_action.png.

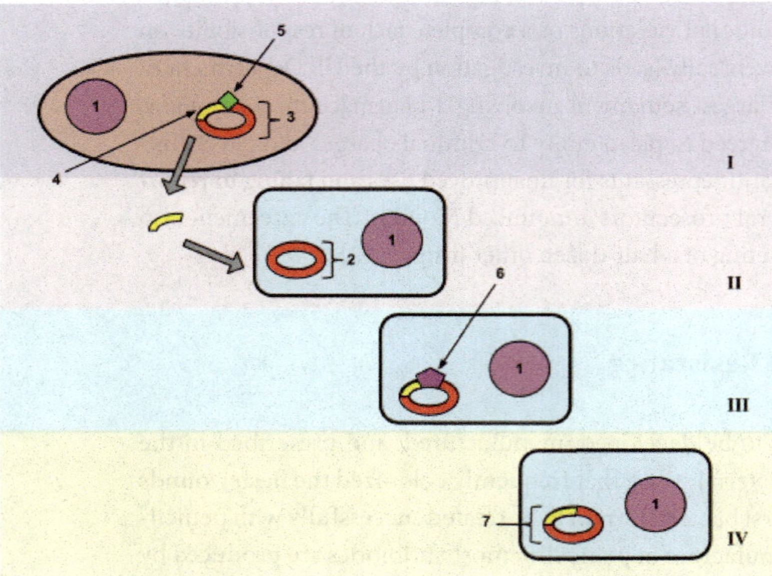

FIGURE 2.10: Bacterial Transformation Conferring Antibiotic Resistance

Step I: The DNA of a bacterial cell is located in the cytoplasm (1), but also in the plasmid, an independent, circular loop of DNA. The gene to be transferred (4) is located on the plasmid of bacterial cell 1 (3), but not on the plasmid of bacterial cell 2 (2). In order to remove the gene from the plasmid of bacterial cell 1, a restriction enzyme (5) is used. The restriction enzyme binds to a specific site on the DNA and "cuts" it, releasing the satisfactory gene. Genes are naturally removed and released into the environment usually after a cell dies and disintegrates.

Step II: Bacterial cell 2 takes up the gene. This integration of genetic material from the environment is an evolutionary tool and is common in bacterial cells.

Step III: The enzyme DNA ligase (6) adds the gene to the plasmid of bacterial cell 2 by forming chemical bonds between the two segments which join them together.

Step IV: The plasmid of bacterial cell 2 now contains the gene from bacterial cell 1 (7). The gene has been transferred from one bacterial cell to another, and transformation is complete.

bacterial targets by interrupting bacterial cell translation (Figure 2.9). Bacterial replication and transcription can also be disrupted by the quinolinones and rifampins, and the sulfanilamides inhibit synthesis of essential metabolites. Each of these mechanisms has been challenged by bacterial gene mutation with resulting bacterial resistance (44, 45). Mutations can occur randomly in the genome, although bacteria most often share fragments of DNA by a mechanism called bacterial transformation that occurs between cells (Figure 2.10). Bacteria are also susceptible to infection by viruses, which can transduce bacteria with novel DNA constructs, thus imparting antibiotic-resistant genes. Acquisition of novel DNA constructs conveys antibiotic resistance through several mechanisms of action, including novel mechanisms of drug destruction, drug efflux, and inhibition of antibiotic binding.

Approximately ten years following penicillin's use as a broad-spectrum antibiotic, substrains of *S. aureus* emerged that were resistant to penicillin's mechanism of action (i.e., inhibition of cell wall synthesis). Subsequently, methicillin was created as an alternative to penicillin, although by the late 1950s and early 1960s, several more substrains of *S. aureus* established themselves as methicillin resistant (46). Vancomycin was then approved by the FDA and became the mainstay of therapy for methicillin-resistant *S. aureus* (MRSA) infection. In 2002, strains of MRSA became resistant to vancomycin (VRSA). The gene cassette that had been acquired by these strains encoded for a protein that replaced the original target on the bacterial cell wall, hence vancomycin lost its target, and the strains became resistant to therapy. Strains of pathogens that cause pneumonia and meningitis, such as *Escherichia coli*, *Klebsiella pneumoniae*, and *Pseudomonas aeruginosa*, have also become resistant to most clinically used antibiotics (47).

As of 2009, morbidity rates associated with MRSA infections rose to 20 percent of the total number of patients initially affected, resulting in approximately nineteen thousand deaths (48). This rate continues to increase, along with the significant expenses and lengthy hospital stays associated with infections. Ten days of an extended hospital stay due to MRSA infection can cost an additional $30,000 per patient, with total US annual expenditures on MRSA infection topping $3 billion per year (49). MRSA is not only acquired in hospitals (HA) but is now a fairly common pathogen found in communities (CA). *S. aureus* is able to colonize environments that tend to be moist and warm, such as athletic facilities and dormitories. Bacterial colonies propagate in towels, dirty clothing, athletic gear, and showers. Increased prevalence of colonies in these locations leads to an increased incidence of infections in younger populations and faster spread among college-aged students.

Methods to develop novel antibiotics focus on bacterial genomics (50). Bacteria employ their natural antibiotics as chemical warfare against rival strains in order to establish territorial dominance. These naturally occurring compounds have provided the basis for traditionally manufactured antibiotics. Bacterial gene cassettes can now be "mined," or sequenced and rearranged in the laboratory, to provide innovative antibiotic compounds. Gene sequences that have remained dormant in parental bacterial strains can be induced to express active compounds when spliced into the genomes of other bacteria. Investigative efforts now look beyond the traditional bacterial host for antibiotic expression to mutualistic organisms that support bacterial growth. Organisms such as fungi and pine beetles provide bacteria an environment in which to live, and in return, bacteria provide these organisms with targeted antibiotic protection against pathogenic bacterial strains. These protective compounds can be isolated and purified to provide protection for humans. Future research will also involve viral systems like bacteriophages and purely synthetic systems, including carbon nanotubes, to develop narrow-spectrum antibiotics and nonbiological platforms for pathogenic bacterial destruction.

f. Drugs of Use and Abuse

Methylphenidate, commonly known as Ritalin®, a drug developed and marketed in 1955 as a remedy for narcolepsy and a "catchall" for other psychiatric disorders, is an excellent example of DTC and its effects on target patient populations. By 2007, 85 percent of the world's consumption of methylphenidate was in the United States, and the therapeutic target was no longer narcolepsy in the older population but attention deficit disorder (ADD) in the juvenile population (51). How did this transition occur? The evolution of Ritalin® as a common drug used to treat attention deficit and hyperactivity in juveniles began with the FDA's approval for its use in children in 1961 (51). During this time, a marketing campaign delivered by Novartis, Inc. (originally the Ciba Company) portrayed Ritalin® as a gentle drug that could be used to treat many, if not all, forms of psychiatric disorders. This portrayal aligned fortuitously with a surge of physician-diagnosed

ADD and attention deficit and hyperactivity disorder (ADHD) in adolescent boys in the 1960s and 1970s (51). Although ADD/ADHD—also known at the time as *minimal brain dysfunction*—was originally diagnosed as a potential childhood disorder in the 1920s, prevalence of this type of diagnosis did not increase until the drug Ritalin® was made available. Marketing advertisements targeted at physicians suggested causation between unruly boys and minimal brain dysfunction. Ritalin® was depicted as the treatment of choice that could reunite frustrated parents and their disruptive children, thus restoring peace at home and in school. Diagnosis and prescriptive treatment skyrocketed in the 1990s. By 2007, approximately 9.5 percent, or 5.4 million children four to seventeen years of age, had been diagnosed with ADD/ADHD, and 66.3 percent of those children were receiving medication treatment for the disorder (51).

Currently, three diagnoses that delineate ADD from ADHD are recognized by the Centers for Disease Control (CDC): predominantly inattentive disorder, predominantly hyperactive-impulsive disorder, and a combined disorder (Table 2.7) (52). Individuals who are diagnosed with predominantly inattentive disorder find difficulty in organizing or finishing tasks, paying attention to detail, or following instructions or conversations. Additionally, they are easily distracted and often forget details of daily routines. Those who are diagnosed with predominantly hyperactive-impulsive disorder struggle to remain focused and physically still. Individuals feel restless and impatient and often act impulsively and interrupt others. A combined diagnosis suggests that both sets of symptoms are equally predominant in the person.

Although the diagnoses for ADD/ADHD primarily address patient overstimulation, most of the pharmacological treatments are central nervous system (CNS) stimulants. The two neuro-chemicals targeted by these medications are the stimulatory neurotransmitters dopamine and norepinephrine (Figure 2.11) (53, 54). Both dopamine and norepinephrine are released from vesicles in presynaptic neurons. Following release into the synaptic cleft, the physical space between the pre- and postsynaptic neurons, both neurotransmitters bind to their respective receptors on the postsynaptic neuronal membranes and elicit stimulatory responses that propagate neural signals further down the excitation pathways. Under normal physiological conditions, these neurotransmitters are eventually removed from the synaptic cleft and are sequestered back into presynaptic vesicles by presynaptic reuptake mechanisms. Ritalin®, a methylphenidate, and

TABLE 2.7: Clinical Categories for ADD and ADHD (original table)

Predominantly inattentive disorder	Individuals have difficulty organizing or finishing tasks, paying attention to details, or following instructions or conversations Individuals are easily distracted and often forget details of daily routines
Predominantly hyperactive-impulsive disorder	Individuals struggle to remain focused and physically still. Individuals feel restless and impatient and often act impulsively and interrupt others
Combined diagnosis	Both sets of symptoms are equally predominant in individuals.

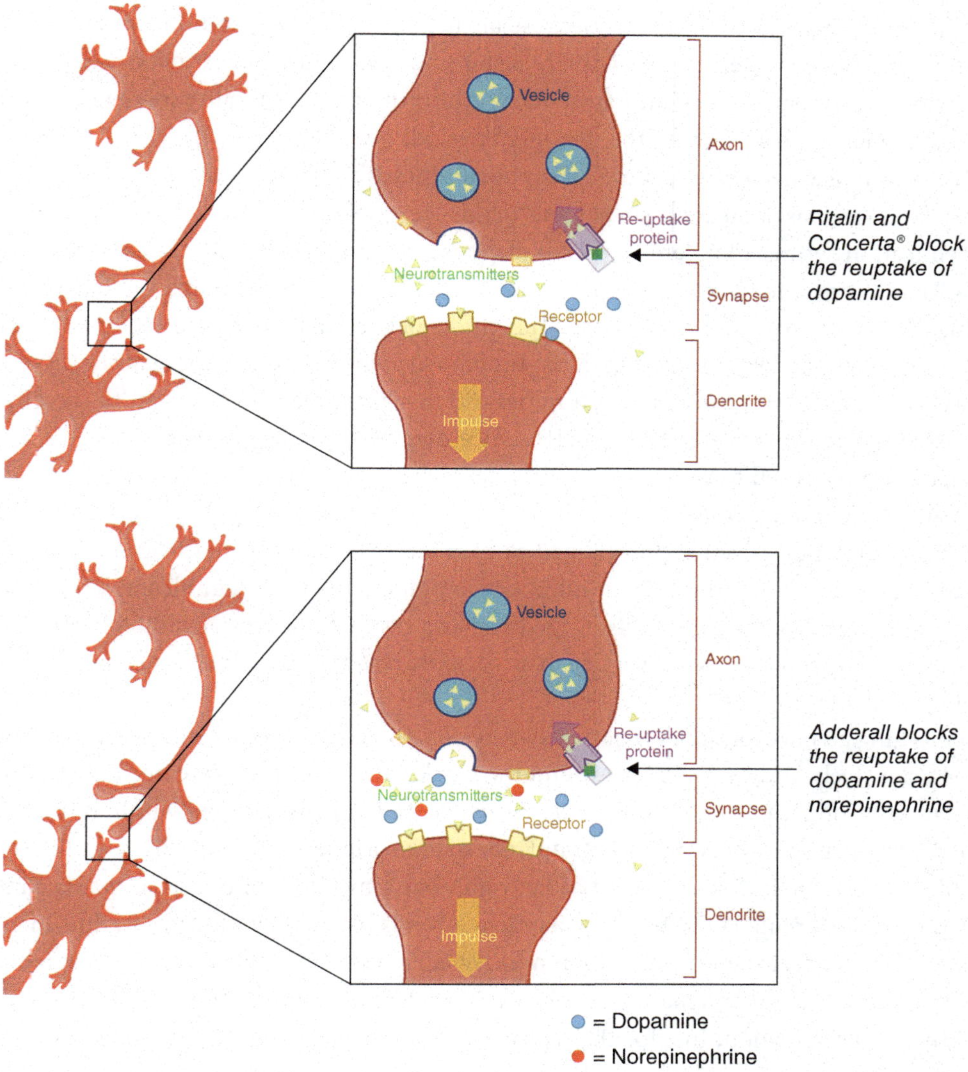

FIGURE 2.11: Mechanisms of Action of Methylphenidates
Sabar / Wikimedia Commons / Copyright in the Public Domain

the extended-release formulation, Concerta®, block the reuptake of dopamine from the synaptic cleft (Figure 2.11) (53, 54). Adderall®, also commonly prescribed for these conditions, appears to block the reuptake of both dopamine and norepinephrine (54). Adderall® has also been postulated to induce additional release of both neurotransmitters into the synaptic cleft, thus increasing their initial concentrations in the cleft. All three medications are typically prescribed at fairly low doses, initially 18 mg twice a day. Increases in dosage are based upon patient-reported responses and physician assessments. Patient reports have suggested that treatment with these and similar medications are effective at controlling inattentiveness and impulsivity. Potential for tolerance to one or all pharmaceuticals is possible, because artificially high concentrations of neurotransmitters within the synaptic cleft can potentially result in down-regulation, or decreases, in relevant receptors on the postsynaptic membranes. As a result, patients may require incrementally higher dosages of medication throughout the treatment period.

As with any pharmaceutical, there are opportunities for abuse. As the prevalence of ADD/ADHD diagnoses has increased, so has the incidence of abuse, particularly in secondary and higher education settings. Students who are diagnosed by physicians and prescribed medication such as Ritalin® are able to sell individual pills for a small fee. Although this is a federal offense, there has been little to discourage abuse. The guidelines for diagnosis also factor into the ease of which students can obtain prescriptions. Individuals who recognize the hallmark symptoms of both ADD and ADHD can manufacture indications fairly easily and present to the physician in ways that suggest either ADD or ADHD. Once obtained, prescriptions can then be used for self-medication or to sell to others. An additional hazard is the way in which these medications may be administered. In contrast to administration of a single-dose pill, medications indicated for abuse are often pulverized and combined so as to increase their potency. Although the mechanisms of action remain the same, the initial concentrations increase significantly, thus flooding the CNS with dangerously high levels of the drug and resultant neurotransmitters. Abuse in this way mimics the actions of street drugs such as cocaine, resulting in similar patterns of abuse and physiological damage.

As emerging global economies require a twenty-four-hour lifestyle and individuals are required to travel extensively and be available around the clock, amphetamines have become increasingly popular to stave off fatigue and enhance performance. Individuals working within business communities have turned to prescription stimulants such as Modafinil®, in addition to Ritalin® and Adderall® (55). Modafinil®, also known as Provigil®, has a mechanism of action that is similar to Ritalin. Dopamine's reuptake in the presynaptic neuron is blocked by Modafinil®, thus increasing the concentration of this neurotransmitter in the synaptic cleft (56). This mechanism results in the propagation of stimulatory signals through neural pathways in the CNS. Initially manufactured and marketed by Cephalon, Inc. for the treatment of narcolepsy and excessive sleepiness due to shift work and sleep apnea, Modafinil® has become popular among the business class as a way to ensure energy and focus. This drug has also been shown, anecdotally, to reduce the effects of jet lag. Although the FDA approved this drug for sleepiness due to third-shift work, Modafinil® was not approved as a treatment for jet lag, and Cephalon, Inc. was required to pay $444 million in fines for promoting Modafinil® and three similar drugs for off-label use (57). The term *cognitive enhancement* is used to define the use of either prescription or nonprescription pharmaceuticals for the purpose of augmenting cognitive function. Currently, there is much debate surrounding the ethics of using performance-enhancing drugs, whether for physical or mental performance. The field of neuroethics has emerged from existing academic disciplines in order to address this controversial practice.

g. Biopharmaceuticals

The desire for biologically targeted approaches to medicine has resulted in the development and manufacture of biopharmaceuticals. In contrast to traditional pharmaceuticals, biopharmaceuticals are biological molecules, most often proteins that are derived from living organisms and function to provide disease-specific therapies. Clinical biomarkers—biological molecules particular to affected patients within a disease state—can be used to identify candidates who have the potential to respond to biopharmaceutical therapies. Biological therapies are developed based upon these particular biological markers, which can also provide ways to assess therapeutic efficacy (58, 59). Two well-known biopharmaceuticals are Enbrel®, a tumor necrosis factor receptor fusion protein used to treat psoriasis and rheumatoid arthritis, manufactured by Amgen, Inc., and Herceptin®, a monoclonal antibody designed for Her2/Neu positive breast cancer patients that is manufactured by Genentech, Inc. Biopharmaceutical molecules can include peptides, proteins, nucleic acids, carbohydrates, or combinations of molecules. Bacteria, mammalian cells, and yeast serve as manufacturing vessels for large-scale production of raw material, followed by isolation and purification procedures that fine-tune the final biological product. Although companies that commercialize pharmaceuticals, biopharmaceuticals, or both must adhere to similar FDA approval pathways, production of biopharmaceuticals is significantly more complicated due to the inherent complexities of the biological products. Unlike simple small-molecule drugs such as aspirin, biopharmaceuticals must maintain specific three-dimensional configurations in order to be biologically active (Figure 2.12) (60). Structure is intimately linked to function

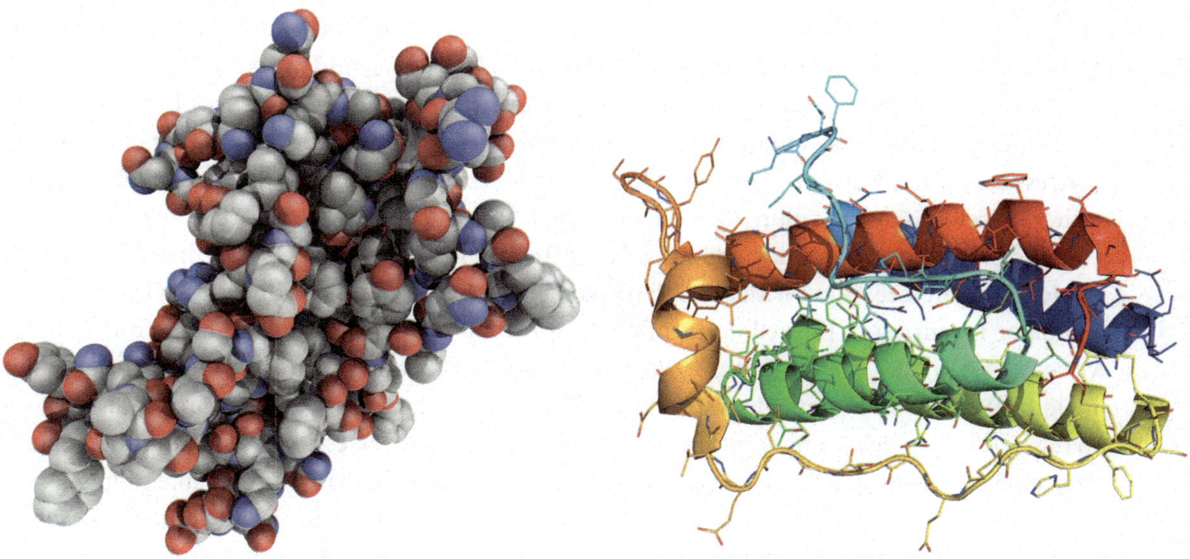

FIGURE 2.12: 3D Structure of Protein Therapeutic: Insulin

FIGURE 2.13: Biopharmaceutical Upstream Processing: Cell Culture

in first-generation biopharmaceuticals, including non-engineered murine-derived monoclonal antibodies and biomimetic replacement proteins, as well as second-generation products, such as genetically engineered fusion proteins. Manufacture of final biopharmaceutical products begins with *upstream processing*, which involves the growth and culture of the biological vessels (i.e., bacteria, mammalian cells, yeast) in which the products will be transcribed and translated (Figure 2.13) (61). Cells are cultured according to their specific guidelines, and once cell densities have been optimized, the cultures are managed so that the biologics can be synthesized. Subsequent methods used to isolate and purify final biological compounds are part of an overall process called *downstream processing*. Once transcribed and translated, biological molecules are either secreted into the culturing media in which the cells are grown or are designated by the cellular genome to be sequestered in vesicles within the cell's membranes. Downstream isolation protocols, therefore, begin with either harvesting of the relevant cell products from the culturing media or rupturing of the cell's membranes in order to extract relevant products from cellular vesicles (Figure 2.14) (61). Both downstream protocols continue with purification steps that include filtration and chromatography procedures that further refine the final products. Because manufacturing relies on biological rather than synthetic processes to produce the therapeutic molecules, all steps must be carried out in sterile environments that are amenable to living systems. Deviations in temperature, pH, and composition of culturing media, as well as alterations in the procedures used for isolation and purification, can lead to the formation of structurally deviant molecules that are biologically inactive. Product-related impurities, including misfolded proteins and protein aggregates, as well as final molecular stabilities, are challenges during the

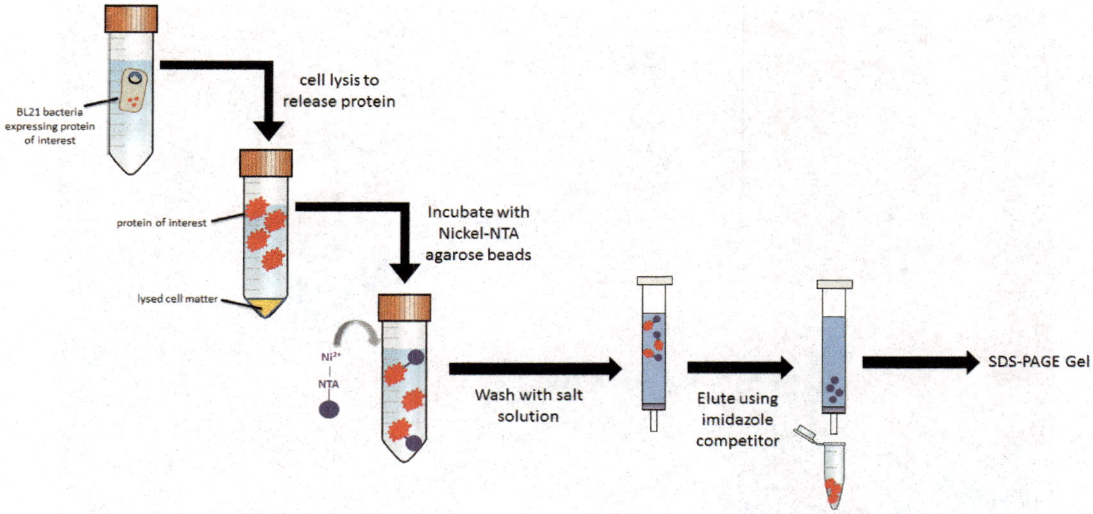

FIGURE 2.14: Biopharmaceutical Downstream Processing: Protein Purification
Copyright © 2014 by RaihaT / Wikimedia Commons, (CC BY-SA 3.0) at http://commons.wikimedia.org/wiki/File:Protein_Purification_MetalBinding.tif.

purification procedures, and each batch of therapeutic must be tested individually for biological activity, purity, and stability. These layers of production complexities, which are innate to biologically derived compounds, lend to their overall clinical effectiveness, albeit at high cost to patients. The wholesale price of Enbrel® is approximately $20,000 per year per patient, and most other products in this market demand similar prices. The global biopharmaceutical industry is currently worth over $145 billion, according to research conducted by BioPlan Associates, and could exceed $167 billion in 2015 (62). Although generic versions of biological drugs, biosimilars or follow-on biologics, are likely to rival brand name products just as generics do in the pharmaceutical industry, there are questions as to whether these products are mechanistically the same (60). Biological systems possess inherent variability even when replicating processes. Manufacturing biosimilars and implementing quality control measures are proving difficult both for manufacturers and the FDA.

Technology on the Edge: Oncolytic Viral Therapy

One of the most significant characteristics of all cancers resides in their abilities to evade patients' immune surveillance systems. The human immune system is designed to recognize invading pathogens such as bacteria and viruses, as well as foreign tissues. Immune cells and molecules are also educated to identify and destroy host cells that have mutated. Mutations can result from chemical carcinogen exposure, radiation and spontaneous genetic alterations that interrupt cellular growth control mechanisms. Normal cells, once responsive to cellular and biochemical signals, become hyperproliferative in the presence of normal growth factors and

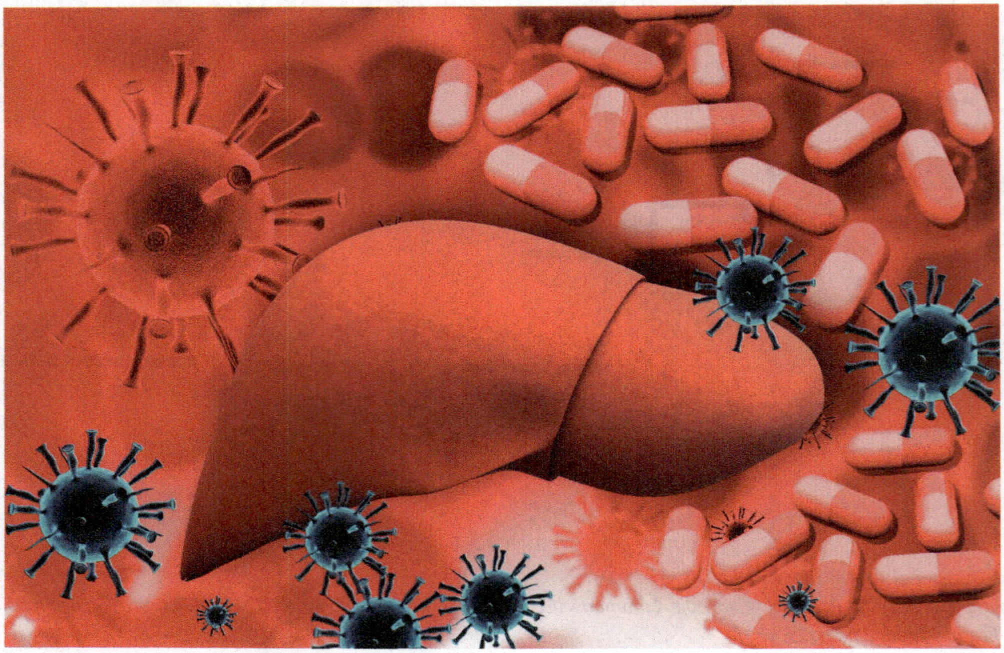

FIGURE 2.15: Concurrent Drug and Viral Therapy for the Treatment of Liver Cancer

are not inhibited by physical cues from surrounding cells and tissues. Cells that have separated from primary tumors and travel systemically are categorized as malignant and are capable of further mutations and continued proliferation.

Approaches to upregulate patients' immune systems in order to overpower and eventually kill malignant cells have been investigated for decades. Observations in the late nineteenth century, anecdotally associating cancer remission in select patients with concurrent viral infections, have led to the search for viral therapies that can treat or even cure cancers. Several types of viruses such as herpes simplex, polio, and vaccinia, have been studied for their anticancer effects. Guo and colleagues have shown that viral-induced cancer cell death occurs by several mechanisms such as immunogenic apoptosis, necrosis, and pyroptosis (63). Gujar and Lee have also shown that viral presentation can override the immune system's inability to recognize cancer-associated antigens and can potentially promote immune cell recognition of these antigens, thus providing specific tumor cell responses (64). Upregulation and maturation of immune cells that are cancer-specific have also been shown to be stimulated by viral presentation (65). Cancer specificity and mode of cancer cell death are critically important for successful clinical outcomes. Genetic manipulations of viral strains that disable normal infective pathways and enable cancer-specific infection and destruction are key determinants of therapeutic viability (66). Additional measures to be considered include regulating viral titres such that patients tolerate viral loads, specific timing of therapeutic infections in order to maximize cancer cell destruction and concurrent viral and drug regimens. Several clinical trials are currently underway to evaluate safety and efficacy of this novel cancer treatment.

End-of-Chapter Questions and Answers

1. What were the reasons for the side effects associated with early pharmaceutical drug formulations?

   Inadequate compound purification and drug dosing
2. How are drugs such as amphetamines, which are modified versions of naturally existing compounds, categorized?

   Semisynthetic drugs
3. In 1962, the Kefauver-Harris Amendment was established. What did this amendment require of pharmaceutical companies?

   Three phases of clinical trials were necessary in order to establish drug safety and efficacy
4. How many compounds are tested as preliminary drug candidates during the development process in a pharmaceutical company?

   Five thousand
5. The drug development process can be divided into five distinct stages. What are these stages?
   a. **Prediscovery**
   b. **Discovery and preclinical**
   c. **Investigational new drug application**
   d. **Clinical trials**
   e. **New drug application**
6. What are the four pharmacokinetic indicators tested for in phase I clinical trials?
   a. **Adsorption**
   b. **Distribution**
   c. **Metabolism**
   d. **Excretion**
7. Which phase of clinical trials tests the safety and efficacy of an investigational new drug?

   Phase I
8. Which phase of clinical trials is designed to be prospective, randomized, placebo controlled, and double blinded?

   Phase III
9. What is the expedited drug-approval process that allows the FDA to collect fees from companies that submit applications for investigational new drugs?

   Prescription Drug User Fee
10. Under the Controlled Substances Act, which classification is given to a pharmaceutical drug that has accepted medical uses, albeit with high potential for abuse?

Schedule II

11. What are the three types of direct-to-consumer (DTC) advertisements that exist for marketing pharmaceutical drugs?
    a. **Product claim**
    b. **Help-seeking**
    c. **Reminder**
12. Antibiotic resistance can occur through bacterial gene mutation and DNA transfer between bacterial cells. What are the two ways that bacterial transfer of genomic fragments most often occur?
    a. **Horizontal gene transfer**
    b. **Cellular conjugation**
13. Which of the following is <u>not</u> a way in which bacteria evade destruction by antibiotics?
    a. **Destruction of the drug target**
    b. **Reduced binding of drug target**
    c. **Chemical modification of the antibiotic**
    d. **Efflux of the antibiotic from the bacterial cell**
14. Which neurotransmitter is affected by the mechanisms of action of methylphenidates such as Ritalin®?
    Dopamine
15. Which stimulant drug works by entering presynaptic neurons and inducing exocytosis of dopamine into the synaptic cleft?
    Adderall®
16. Which protein characteristic must be maintained in order for a therapeutic protein to remain biologically active following biopharmaceutical processing and manufacture?
    3-D conformation
17. Biopharmaceutical manufacturing involves two distinct processes. What are these two processes?
    a. **Upstream processing**
    b. **Downstream processing**
18. Which type of pharmaceutical advertising must directly state major risk information and must direct viewers to other sources of information?
    Broadcast advertising
19. What is the name of the biopharmaceutical therapy that acts as a TNF receptor agonist for the treatment of rheumatoid arthritis?
    Enbrel®
20. What types of biological cells are used for producing biopharmaceutical proteins?
    Mammalian
    Bacterial
    Yeast

Works Cited

1. Maxwell, Jane Carlisle. "The Prescription Drug Epidemic in the United States: A Perfect Storm." *Drug and Alcohol Review* 30.3 (2011): 264–70.
2. Garrett, Nancy, and E. Mary Martini. "The Boomers Are Coming: A Total Cost of Care Model of the Impact of Population Aging on the Cost of Chronic Conditions in the United States." *Disease Management* 10.2 (2007): 51–60.
3. Katiyar, Chandrakant, Satyajyoti Kanjilal, Arun Gupta, and Shefali Katiyar. "Drug Discovery from Plant Sources: An Integrated Approach." *AYU (An International Quarterly Journal of Research in Ayurveda)* 33.1 (2012): 10.
4. Davies, D. M. "2000 Years of Adverse Drug Reactions." *Adverse Drug Reaction Bulletin* 199.1 (1999): 759–62.
5. Rosinsky, Natalie M. *The Story of Pharmaceuticals: How They Changed the World*. Mankato, MN: Compass Point, 2010.
6. Hunting, P. "The Worshipful Society of Apothecaries of London." *Postgraduate Medical Journal* 80.939 (2004): 41–44.
7. Dailey, John W. "Pharmaceutical Science in the 16th and 17th Centuries." *Encyclopaedia Britannica Online*. Encyclopaedia Britannica.
8. Toledo-Pereyra, Luis H. "Louis Pasteur Surgical Revolution." *Journal of Investigative Surgery* 22.2 (2009): 82–87.
9. Toledo-Pereyra, Luis H. "Joseph Lister's Surgical Revolution." *Journal of Investigative Surgery* 23.5 (2010): 241–43.
10. Rep. N.p.: n.p., n.d. *Healthcare at a Glance; OECD Indicators*. OECD Publishing, 2013.
11. "Penicillin Production through Deep-tank Fermentation." *National Historic Chemical Landmark*. American Chemical Society, 2008.
12. Drews, J. "Drug Discovery: A Historical Perspective." *Science* 287.5460 (2000): 1960–964.
13. Liao, Wen-Ling, and Fuu-Jen Tsai. "Personalized Medicine: A Paradigm Shift in Healthcare." *BioMedicine* 3.2 (2013): 66–72.
14. Lahlou, Mouhssen. "The Success of Natural Products in Drug Discovery." *Pharmacology & Pharmacy* 04.03 (2013): 17–31.
15. Appendino, Giovanni, Alberto Minassi, and Orazio Taqlialatela-Scafati. "Recreational Drug Discovery: Natural Products as Lead Structures for the Synthesis of Smart Drugs." *Natural Products Report* 31 (2014): 880–904.
16. Stephens, Trent D., and Rock Brynner. *Dark Remedy: The Impact of Thalidomide and Its Revival as a Vital Medicine*. Cambridge, MA: Perseus Pub., 2001.
17. "About FDA." *Overviews on FDA History*.
18. Act of October 10, 1962 (Drug Amendments Act of 1962), Public Law 87-781, 76 STAT 780, Which Amended the Federal Food, Drug, and Cosmetic Act to Assure the Safety, Effectiveness, and Reliability of Drugs, Authorize Standardization of Drug Names, and Clarify and Strengthen Existing Inspection Authority, 10/10/1962.

19. Boehm, Garth, Lixin Yao, Liang Han, and Qiang Zheng. "Development of the Generic Drug Industry in the US after the Hatch-Waxman Act of 1984." *Acta Pharmaceutica Sinica B* 3.5 (2013): 297–311.
20. Kesselheim, A. S., and J. J. Darrow. *Drug Development and FDA Approval, 1938–2013.* 370.26 (2014).
21. Hughes, J. P., S. Rees, S. B. Kalindjian, and K. L. Philpott. "Principles of Early Drug Discovery." *British Journal of Pharmacology* 162.6 (2011): 1239–249.
22. *Drug Discovery and Development.* Publication. Washington, DC: Innovation.org, 2007. Print.
23. Umscheid, Craig, David Margolis, and Craig Grossman. "Key Concepts of Clinical Trials: A Narrative Review." *Postgraduate Medicine* 123.5 (2011): 194–204.
24. Rubin, Eric H., and D. Gary Gilliland. "Drug Development and Clinical Trials—the Path to an Approved Cancer Drug." *Nature Reviews Clinical Oncology* (2012).
25. "For Consumers." *FDA 101: Clinical Trials and Institutional Review Boards.* FDA, 2009.
26. *The FDA's Drug Review Process: Ensuring Drugs Are Safe and Effective.* FDA, 2014.
27. Collier, R. "Rapidly Rising Clinical Trial Costs Worry Researchers." *Canadian Medical Association Journal* 180.3 (2009): 277–78.
28. Kramer, Judith M., and Kevin A. Schulman. *Transforming the Economics of Clinical Trials.* Working paper. N.p.: National Academy of Sciences, 2012.
29. "For Industry." *Prescription Drug User Fee Act (PDUFA).* FDA.
30. Franson, T. R. "Has the FDA Amendments Act of 2007 Impaired Drug Development?" *Clinical Pharmacology and Therapeutics* 89.2 (2011): 169–71.
31. "Regulatory Information." *Controlled Substances Act.* FDA, 2009.
32. 2014. "DEA/Drug Scheduling." *DEA/Drug Scheduling.*
33. Greene, Jeremy A., and Scott H. Podolsky. "Reform, Regulation, and Pharmaceuticals—The Kefauver-Harris Amendments at 50." *New England Journal of Medicine*367.16 (2012): 1481–483.
34. "An Assessment of Direct-to-Consumer Advertising of Prescription Drugs Perspective." *Clinical Pharmacology & Therapeutics* 86 (2007): 357–60.
35. Berndt, Ernst R. "To Inform or Persuade? Direct-to-Consumer Advertising of Prescription Drugs." *New England Journal of Medicine* 352.4 (2005): 325–28.
36. Rubin, Lawrence C. "Merchandising Madness: Pills, Promises, and Better Living Through Chemistry." *Journal of Popular Culture* 38.2 (2004): 369–83.
37. Ventola, C. L., MS. "Direct-to-Consumer Pharmaceutical Advertising: Therapeutic or Toxic?» *Pharmacy & Therapeutics* 36.10 (2011): 669+.
38. "For Consumers." *Keeping Watch Over Direct-to-Consumer Ads.* FDA, 19 Aug. 2013.
39. Fugh-Berman, Adriane, and Shahram Ahari. "Following the Script: How Drug Reps Make Friends and Influence Doctors." *PLoS Medicine* 4.4 (2007): E150.
40. *Code on Interactions with Healthcare Professionals.* Rep. Phrma.org, 2008. www.phrma.org/sites/default/files/pdf/phrma_marketing_code_2008.pdf.

41. "Drugs." *Advertising.* FDA, 16 June 2014. www.fda.gov/Drugs/GuidanceComplianceRegulatoryInformation/Guidances/ucm064956.htm.
42. Bass, Alison. *Side Effects: A Prosecutor, a Whistleblower, and a Bestselling Antidepressant on Trial.* Chapel Hill, NC: Algonquin of Chapel Hill, 2008.
43. American Chemical Society. *Discovery and Development of Penicillin.* London: American Chemical Society, 1999.
44. Talaro, Kathleen P., and Barry Chess. *Foundations in Microbiology.* New York: McGraw-Hill, 2012.
45. *Centers for Disease Control and Prevention.* Centers for Disease Control and Prevention, 18 Dec. 2013. www.cdc.gov/getsmart/antibiotic-use/antibiotic-resistance-faqs.html.
46. Aminov, Rustam I. "A Brief History of the Antibiotic Era: Lessons Learned and Challenges for the Future." *Frontiers in Microbiology* 1 (2010).
47. Chambers, Henry F., and Frank R. DeLeo. "Waves of Resistance: Staphylococcus Aureus in the Antibiotic Era." *Nature Reviews Microbiology* 7 (2009): 629–41.
48. "Methicillin-resistant *Staphylococcus aureus* (MRSA) Infections." *MRSA Tracking.* Centers for Disease Control and Prevention. www.cdc.gov/mrsa/tracking/index.html.
49. Rosner, Andrew J. "The Costs and Consequences of Methicillin-Resistant Staphylococcus Aureus Infection Treatments in Canada." *Can J Infect Dis Med Microbiol* 15.4 (2004): 213–20.
50. Rossolini, Gian M. "Coping with Antibiotic Resistance: Contributions from Genomics." *Genome Medicine* 2.15 (2010).
51. Rubin, Lawrence C. "Merchandising Madness: Pills, Promises, and Better Living Through Chemistry." *Journal of Popular Culture* 38.2 (2004).
52. *ADHD Symptoms and Diagnosis.* Centers for Disease Control and Prevention. www.cdc.gov/ncbddd/adhd/diagnosis.html.
53. Onyebeke, Chukwuma. "Study Drugs and Neural Enhancers: Science and Controversy." *Yale Scientific* 9 May 2012.
54. Josephine, Elia, MD. "Attention Deficit/Hyperactivity Disorder Pharmacotherapy." *Psychiatry* 2.1 (2005): 27–35.
55. Szalavitz, Maia. "Popping Smart Pills: The Case for Cognitive Enhancement." *Time* 6 Jan. 2009.
56. Turner, Danielle C., Trevor W. Robbins, Luke Clark, Adam A. Aron, Jonathan Dowson, and Barbara J. Sahakian. "Cognitive Enhancing Effects of Modafinil in Healthy Volunteers." *Psychopharmacology* 165.3 (203): 260–69.
57. Pringle, Evelyn. "The Rise and Fall of Provigil." *Salient News* 21 Sept. 2010.
58. Rader, Ronald A. "(Re)defining Biopharmaceutical." Editorial. *Nature Biotechnology* July 2008: 743–51.
59. Baker, Monya. "In Biomarkers We Trust?" *Nature Biotechnology* 23.3 (2005): 297–304.
60. Gottlieb, Scott. "Biosimilars: Policy, Clinical, and Regulatory Considerations." *Am J Health-Syst Pharm* 65.6 (2008): 52–58.

61. Jameel, Feroz, and Susan Hershenson. *Formulation and Process Development Strategies for Manufacturing Biopharmaceuticals.* Hoboken, NJ: John Wiley & Sons, 2010.
62. Langer, Eric S. "Biomanufacturing Outsourcing Outlook—Industry Optimism Is on the Rise for 2012.—BioPharm International." BioPharm International, 1 Feb. 2012. www.biopharminternational.com/biopharm/article/articleDetail.jsp?id=757935&sk=&date=&pageID=3.
63. Guo, Zong Sheng, Zuqiang Liu, and David L. Bartlett. "Oncolytic Immunotherapy: Dying the Right Way Is a Key to Eliciting Potent Antitumor Immunity." *Front. Oncol. Frontiers in Oncology* 4 (2014): n. pag. Web.
64. Gujar, Shashi A., and Patrick W. K. Lee. "Oncolytic Virus-Mediated Reversal of Impaired Tumor Antigen Presentation." *Front. Oncol. Frontiers in Oncology* 4 (2014): n. pag. Web.
65. Moehler, Markus, Katrin Goepfert, Bernd Heinrich, Caroline J. Breitbach, Maike Delic, Peter Robert Galle, and Jean Rommelaere. "Oncolytic Virotherapy as Emerging Immunotherapeutic Modality: Potential of Parvovirus H-1." *Front. Oncol. Frontiers in Oncology* 4 (2014): n. pag. Web.
66. Schirrmacher, Volker, and Philippe Fournier. "Multimodal Cancer Therapy Involving Oncolytic Newcastle Disease Virus, Autologous Immune Cells, and Bi-Specific Antibodies." *Front. Oncol. Frontiers in Oncology* 4 (2014): n. pag. Web.

CHAPTER 3
Organ Transplantation

a. Significant Milestones in Transplantation Medicine

On average, in the United States, a hundred thousand patients each year are listed on the organ transplant waiting list, as documented by the United Organ Sharing Network (1). The number of potential organ donors, however, pales in comparison, and therein resides the problem. Approximately thirty thousand transplants are carried out each year in the United States (1). Due to the large discrepancy between the number of available organs and the number of patients in need, thousands of patients die each year while awaiting transplants. Effective trauma management and the emergence of successful therapies for cardiovascular and pulmonary disorders and resulting increases in life expectancies have led to decreases in the number of potential organ donations. Despite this challenge, however, organ transplantation remains the gold standard for replacing nonfunctional or diseased organs and tissues.

The first successful transplant of a nonvascularized tissue—the cornea—occurred over a hundred years ago in Austria (2). Approximately fifty years later, the first living kidney donation and transplant, between identical twins, were conducted by Drs. Joseph Murray and David Hume at Brigham Hospital in Boston, Massachusetts, and in 1962, the first successful cadaveric (deceased) kidney transplant was performed by these same surgeons (Figure 3.1) (3). Transplantations of heart, liver, and lung soon followed, and in 1983, the FDA approved cyclosporine, the most successful antirejection medication developed to date (4).

Additional milestones have been realized with the establishment of the National Organ Transplant Act in 1984, which launched a nationwide computer registry operated by the United Network for Organ Sharing. The act also authorized financial support for organ procurement organizations (OPOs) and prohibited the buying and selling of organs in the United States (5).

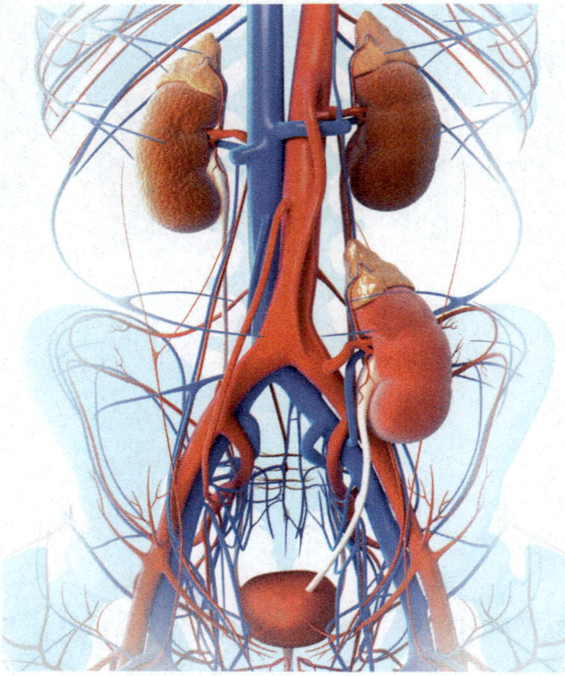

FIGURE 3.1: Kidney Transplantation
Copyright © by Depositphotos / pixologic.

Additionally, required request laws were created to ensure that hospitals develop policies to identify patients as potential donors and approach families about organ donations (6).

The establishment of these guidelines has guaranteed that all hospitals are required by law to have a *required referral* system in place. Under this system, the hospitals must notify the local OPO of all patient deaths. If the OPO determines that organ and/or tissue donation is appropriate in a particular case, it has a representative contact the deceased patient's family to offer them the option of donating their loved one's organs and tissues. By signing a Uniform Donor Card, an individual indicates his or her wish to be a donor. However, at the time of death, the person's next-of-kin will still be asked to sign a consent form for donation. It is estimated that about 35 percent of potential donors never become donors because family members refuse to give consent (7). All costs related to the donation of organs and tissues are paid for by the donor program. Aside from living kidney donations, acceptable organ donors are those whose brain function has ceased permanently, but whose heart and lungs continue to function with the use of ventilators. Organ transplant recipients are selected on the basis of medical urgency, as well as compatibility of body size and blood chemistries—not race, sex, or creed. Advances in surgical techniques and organ preservation and the development of more effective drugs to prevent rejection have improved the success rates of all types of organ and tissue transplants (Table 3.1) (8).

TABLE 3.1: Species-Defined Types of Transplanted Organs (original table)

Type	Description	Immunosuppressants Required?
Autologous	Transplantation of cells/tissues/organs within the same individual	No
Allogeneic	Transplantation of cells/tissues/organ between different individuals of the same species	Yes
Xenogeneic	Transplantation of cells/tissues/organ between individuals of different species	Yes

b. Tissue Sources

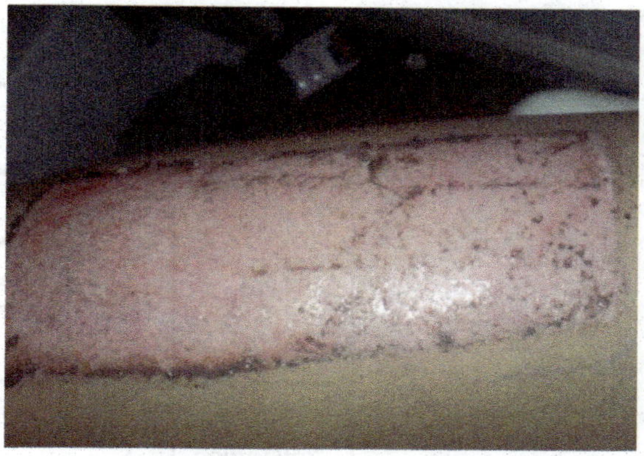

FIGURE 3.2: Autologous Skin Graft
Kevin308 / Wikimedia Commons / Copyright in the Public Domain.

Allogeneic transplantation—transplantation of organs and tissues between individuals—has been the customary practice in modern medicine for approximately fifty years (9). In fact, evidence has shown that allogeneic transplantation was practiced thousands of years ago by the ancient Egyptians. In addition to allogeneic transplantation, autologous and xenogeneic transplants are options. With autologous tissue transplantation, tissues are harvested and transplanted within the same patient, such as with skin grafting of burn patients (Figure 3.2) (10). Partial and full thickness skin grafts can be harvested from unaffected areas on the body and transferred to burn sites. Autologous skin grafting has proven very successful for effective recovery from second- and third-degree burns as long as there is enough healthy skin from which to harvest.

Xenogeneic transplantation refers to transplantation of organs and tissues between members of different species (11). One of the most high-profile cases involved the transplantation of a baboon heart into Baby Fae, an infant suffering from a fatal congenital heart defect, in 1984 (12, 13). Baby Fae lived for twenty days following transplantation and succumbed to immunological rejection reactions elicited by her immune system against the foreign organ. In spite of the immunological intolerance by Baby Fae's immune system, the concept of xenogeneic transplantation still remains viable because it provides a solution to the lack of available organs for donation. Strategies to inhibit rejection responses while supporting patient health during xenogeneic transfer must be accomplished before future successful cases can be attempted. Immunological intolerance of donated tissue is prevalent not only with xenogeneic transplants, but also with allogeneic transplants, albeit rejection reactions can be somewhat controlled in the latter case.

c. Immunology of Transplantation and Immunosuppression

Organ transplant rejection is caused by an immunological attack of transplanted allogeneic and xenogeneic organs by the recipient's immune system. Approximately three immunological pathways are involved in the rejection responses that lead to death of transplanted tissues. As a result, donor and recipient are screened for compatibility using several diagnostic tests.

The first screening method is the matching of blood types between donor and recipient (Table 3.2) (14). Red blood cells (RBCs) carry antigenic proteins—proteins that can potentially cause immune reactions—on their surfaces. These proteins, and hence the RBCs that present them, are designated A, B, and AB; if no antigenic protein is present on the surface, then blood

TABLE 3.2: ABO Blood Typing

	Group A	Group B	Group AB	Group O
Red blood cell type	A	B	AB	O
Antibodies in Plasma	Anti-B	Anti-A	None	Anti-A and Anti-B
Antigens in Red Blood Cell	A antigen	B antigen	A and B antigens	None

Source: InvictaHOG / Wikimedia Commons / Copyright in the Public Domain.

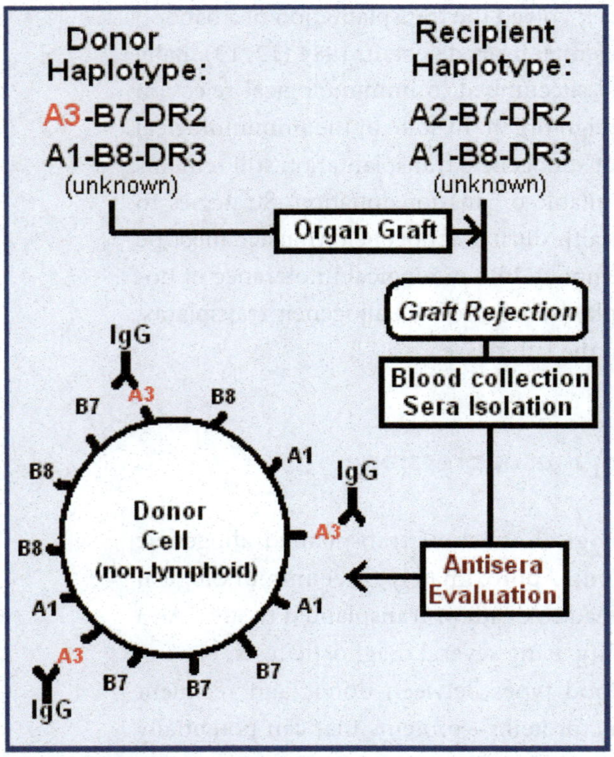

FIGURE 3.3: An Example of Donor–Recipient HLA Mismatch
Pdeitiker / Wikimedia Commons / Copyright in the Public Domain.

type is defined as O. Individuals with blood type A produce antibodies directed against blood type B. Individuals with blood type B produce antibodies against blood type A. Those with blood type AB produce no antibodies, and patients with blood type O produce both antibodies. Antibodies against a particular RBC antigen bind to the antigens located on the RBC surfaces and cause agglutination of the blood cells. In transplanted organs, this results in obstruction of tissue capillaries and tissue death.

The second screening method involves identification of specific immunological cell markers called human leukocyte antigens (HLA) that are located on the surfaces of all nucleated cells (Figure 3.3) (15, 16). These proteins are individualized, and their specific genotype and phenotype can only be truly matched between identical twins. Measures are taken to match donors and recipients as closely as possible, although lifelong immunosuppressive therapy is required by the transplant recipient. In 1967, the discovery of HLA—called major histocompatibility complex (MHC) in nonhuman species—shed light on the mechanisms behind organ transplant rejection. Genes for these cell surface antigens are located on the short arm of chromosome 6 and total two hundred genes in all. Three classes of HLA—I, II, and III—exist and are distinguished based upon tissue distribution, structure, and function. Genes for HLA I and II encode for cell surface antigens of nucleated cells and immune cells, respectively. Genes for HLA III encode for immune complement components. Combinations of HLA I, II, and III are inherited in a Mendelian-dominant manner and are called immunological haplotypes.

Various diagnostic tests can be used to determine tissue compatibility between potential donors and recipients. Blood typing is assessed by a standard ABO blood test, in which blood samples from donor and recipient are mixed and cellular aggregation, or hemagglutination, is observed (14). HLA reactivity and incompatibility can be ascertained by observing and quantifying white blood cell mitosis following introduction of donor leukocytes to recipient leukocytes in a mixed leukocyte culture (16). Serological cross-matching, used for both HLA and ABO matching, establishes reactivity by quantifying agglutination and hemagglutination reactions, respectively. The third diagnostic test, DNA typing of donor and recipient HLA genomes, is most often used for attaining rapid results (17). This particular method results in definitive allelic mismatch data.

Rejection of transferred allogeneic organs is possible at three distinct phases following transplantation (Table 3.3) (18, 19, 20). The most immediate rejection response elicited by the recipient is a hyperacute rejection response that occurs within the first forty-eight hours posttransplantation. The immunological mode of rejection is antibody mediated, resulting from the presence of preexisting recipient antibodies against donor endothelial cells. Following binding of antibodies to donor endothelial cells located within the transplanted tissues, activation of the complement and coagulation cascades results in rapid donor tissue death (Figure 3.4). This type of reaction is irreversible and is characterized by immediate fever and anuria. Hyperacute reactions are prevalent in xenogeneic, or cross-species, transplantations.

TABLE 3.3: Rejection Reactions (original table)

Type of Rejection	Time Frame	Immune Response
Hyperacute rejection	Immediate: 48 hours posttransplantation	Preexisting recipient antibodies against donor endothelial cells
Acute rejection	6 to 90 days posttransplantation	Recipient T cells directly target donor tissue via HLA I binding/preexisting recipient antibodies against donor endothelial cells
Chronic rejection	90 days and beyond transplantation	T cells indirectly target donor tissue via HLA II binding/generation of antibodies via HLA II presentation of donor antigens

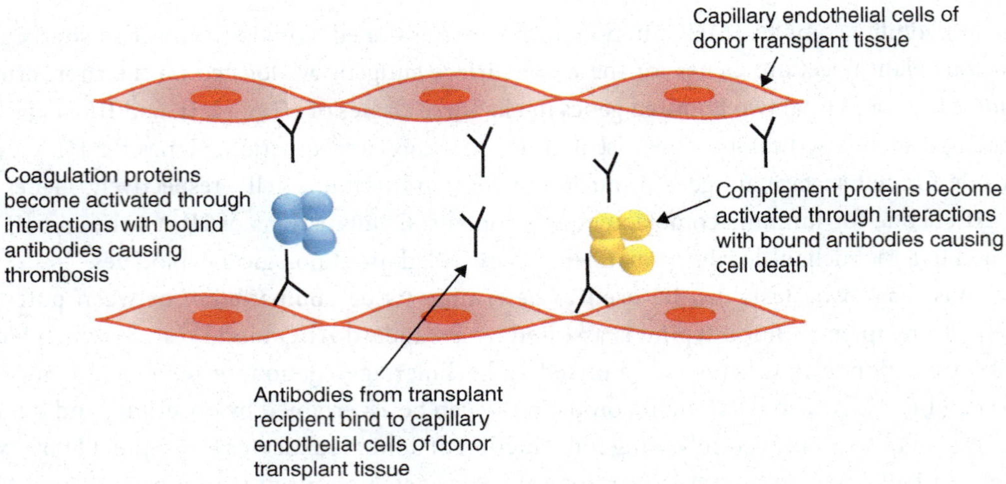

FIGURE 3.4: Hyperacute Immune Reaction (original schematic)

Within six to ninety days posttransplantation, a second type of immunological response, acute rejection, is possible. Both cellular and antibody-mediated reactions contribute to rejection during this phase. Recipient white blood cells, T lymphocytes, target donor tissues directly. T helper cells that are capable of secreting chemoattractant and proinflammatory cytokines, molecules involved in immune cell recruitment and activation, provide immunological support for T cytotoxic cell populations. T cytotoxic cells actively destroy cellular targets by binding to both HLA I antigens and costimulatory moieties on donor tissue cells and secreting substances such as perforin that puncture donor cell membranes (Figure 3.5). Antidonor antibodies against donor endothelial cells

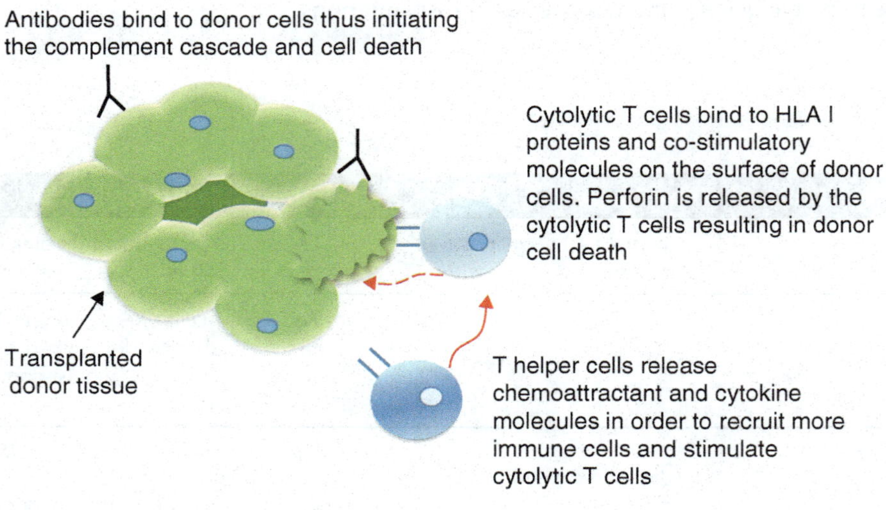

FIGURE 3.5: Acute Rejection Response (original schematic)

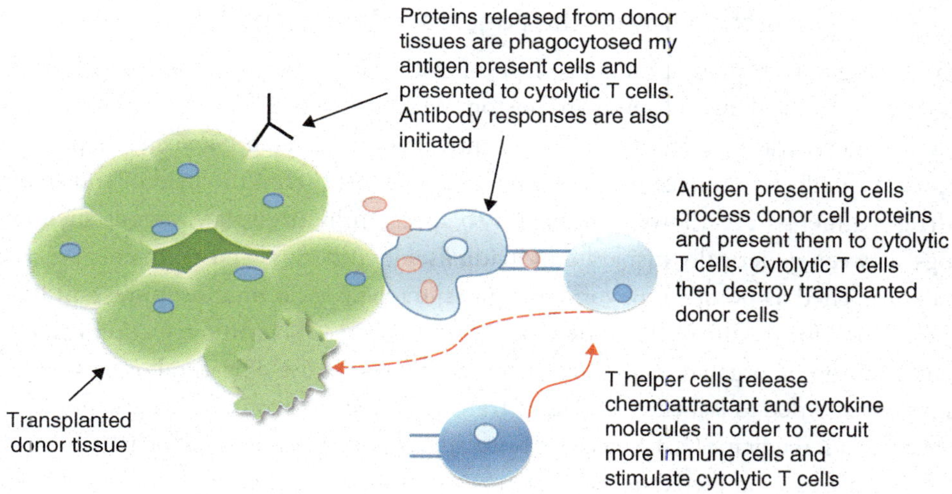

FIGURE 3.6: Chronic Rejection

cause further tissue damage by binding to donor cells and eliciting the complement cascade, which advances the rejection reaction.

The third type of immune response, long-term chronic tissue rejection, which may be subdued by immunosuppressant drugs, may present months to years posttransplantation. This response is considered an indirect immune response by the recipient because it requires antigen processing and presentation to the immune system in order to elicit a response. The basis for rejection is cellular as well as antibody driven and results in transplant death, fibrosis, and scarring. Proteins that are produced by the cells of donated tissues are engulfed by white blood cells such as macrophages, processed, and then displayed on the macrophage surface along with HLA II proteins (Figure 3.6). Presentation in this way stimulates both cellular and antibody responses by the recipient, and if additional costimulatory and supportive molecules are present, then rejection of the donated tissues occurs.

Immunosuppressant drugs such as cyclosporine, which was approved for use in transplant patients in 1983, have significantly improved the functional health of transplanted organs by reducing the activity of T cells, thus reducing the chances of their stimulation and ultimate destruction of donated tissues.

An additional immunological reaction that can be treated using immunosuppressant therapy is graft versus host disease (21). This particular phenomenon involves stimulation of donated bone marrow cells following bone marrow transplantation. Donated bone marrow cells become activated against the recipient's cells and destroy cells and tissues systemically. Patients, who are required to take several different immunosuppressant medications daily, remain on immunosuppressant therapies for their entire lives. These medications are not only expensive—$2,500 per month—but also inhibit patients' healthy immune responses to viral and bacterial infections (22). Patients most often succumb to opportunistic infections and cancer rather than organ failure when following this type of drug regimen.

Technology on the Edge: Face Transplantation

Modern-day organ transplantation began in 1962 with the living donation of a kidney between identical twins at Brigham Hospital in Boston, Massachusetts. Immunological rejection by the organ recipient was avoided due to the identical genetic match between the pair of twins. Development of effective immunosuppressant drugs has furthered the field of transplantation. These drug regimens have allowed physicians to successfully transplant organs such as hearts and lungs between genetically disparate individuals, albeit donors and recipients are as closely genetically matched as possible. Beyond acute tissue rejection, immunosuppressant drug regimens, prescribed for the life of the patient, successfully prevent chronic rejection responses.

Up until this point, organ donation has strictly involved the use of visceral organs. In 2005, the face transplant afforded to Isabelle Dinoire, by transplantation surgeons in Amiens, France, revolutionized transplant medicine (23). Since that time, approximately twenty-eight partial and full facial transplants have been completed in the United States and abroad (24, 25). Face transplantation is an option for patients who have facial deformities due to traumatic facial injuries or cancer and the like. Qualified patients must match—as closely as possible—both the donor's blood type as well as HLA profiles. Anastomosis of the vasculature is critical, and many patients experience episodes of mild-to-moderate acute rejection post-transplantation (26). Aesthetic outcomes vary, although a majority of patients physically present characteristics of both donor and recipient. Although in most cases recipients maintain their underlying skeletal structures, overlying soft tissues from the donor

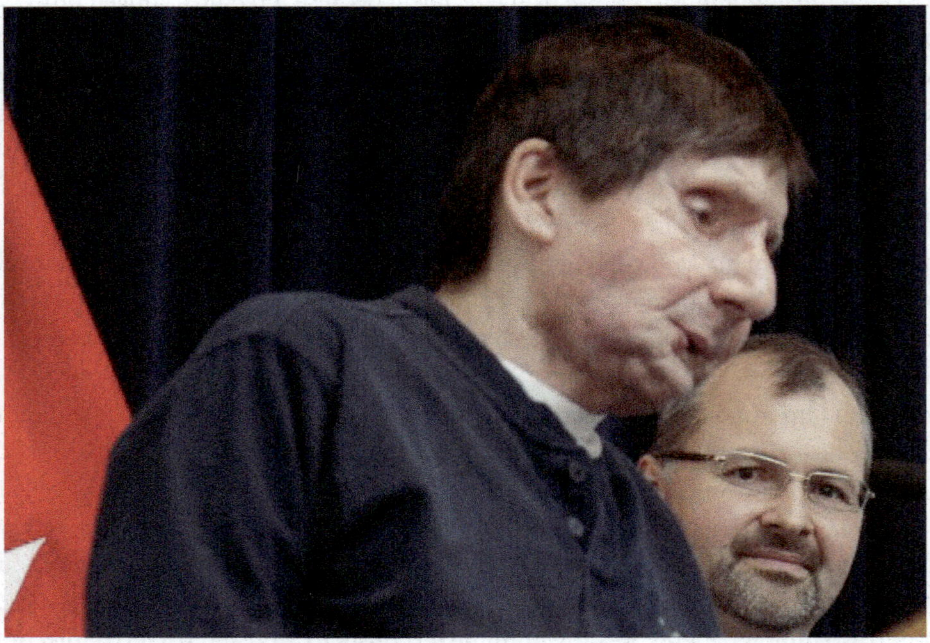

FIGURE 3.7: Jim Maki, facial transplant recipient
David Kamm / U.S. Army / Copyright in the Public Domain.

morphologically change the recipient's appearance to some degree. Long-term psychological effects on recipients are still unknown; however, transplants can, in some instances, restore functional aspects of the nasal and oral cavities, which allow considerable independence for many recipients. Post-transplant revision surgeries and overall expenses continue to prevent many patients from selecting facial transplantation. Significant advances in transplant immunology and microsurgical techniques have the potential to control costs and patient outcomes and may provide more patients the opportunities to choose this revolutionary treatment.

End-of-Chapter Questions and Answers

1. What is the name of the most successful immunosuppressant drug to date that was approved for use by the FDA in 1983 and was developed to prevent organ transplant rejection?
   Cyclosporine
2. Under the guidelines of the required referral system, which party is responsible for determining whether organ/tissue donation is appropriate in each patient case?
   The organ procurement organization
3. What is the classification of cells, tissues, and/or organs that are transplanted between individuals from the same species?
   Allogeneic transplantation
4. What is the classification of cells and/or tissues that are transplanted within the same individual?
   Autologous transplantation
5. What is the classification of cells, tissues, and/or organs that are transplanted between individuals of different species?
   Xenogeneic transplantation
6. Patients who have type O blood may be candidates to donate cells, tissues, and/or organs to patients having which blood types?
   They may donate to patients of any blood type: A, B, AB, or O
7. Which rejection reaction involves recognition of donor tissues by recipient preformed antibodies?
   Hyperacute rejection
8. Which rejection reaction is initiated by donor HLA I presentation to recipient T cells?
   Acute rejection
9. Hyperacute rejection responses are most likely to occur in which type of species-related organ transplantation?
   Xenogeneic transplantation
10. The immunosuppressant drug cyclosporine was designed to prevent which type of rejection reaction?
    Chronic transplant rejection

Works Cited

1. "UNOS DonateLife." *UNOS*. www.unos.org/.
2. Armitage, W. J., A. B. Tullo, and D. F. P. Larkin. "The First Successful Full-thickness Corneal Transplant: A Commentary on Eduard Zirm's Landmark Paper of 1906." *British Journal of Ophthalmology* 90.10 (2006): 1222–223.
3. Tilney, Nicholas L. "Renal Transplantation between Identical Twins: A Review." *World Journal of Surgery* 10.3 (1986): 381–88.
4. Tedesco, Dustin, and Lucas Haragsim. "Cyclosporine: A Review." *Journal of Transplantation* (2012).
5. "OPTN: Organ Procurement and Transplantation Network." *OPTN: Organ Procurement and Transplantation Network*. http://optn.transplant.hrsa.gov/policiesAndBylaws/nota.asp.
6. "Legislation and Policy." *Organdonor.gov*. www.organdonor.gov/legislation/.
7. Branco, B. C., K. Inaba, L. Lam, A. Salim, G. Barmparas, P. G. Teixeira, P. Talvinq, and D. Demetriades. "Donor Conversion and Procurement Failure: The Fate of Our Potential Organ Donors." *World Journal of Surgery* 35.2 (2011): 440–45.
8. Lee, R.-A., and S. Gabardi. "Current Trends in Immunosuppressive Therapies for Renal Transplant Recipients." *American Journal of Health-System Pharmacy* 69.22 (2012): 1961–975.
9. Tobin, G. R., W. C. Breidenbach, S. T. Ildstad, M. M. Marvin, J. F. Buell, and K. V. Ravindra. "The History of Human Composite Tissue Allotransplantation." *Transplantation Proceedings* 41.2 (2009): 466–71.
10. Shimizu, Ruka, and Kazuo Kishi. "Skin Graft." *Plastic Surgery International* (2012).
11. Yamada, Kazuhiko, Adam Griesemer, and Masayoshi Okumi. "Pigs as Xenogeneic Donors." *Transplantation Reviews* 19.3 (2005): 164–77.
12. Altman, Lawrence K. "Baby Fae, Who Received a Heart from Baboon, Dies after 20 Days." *New York Times*, 16 Nov. 1984.
13. Shaefer, Richard A. "Chapter 2: Baby Fae." *LLUMC Legacy: Daring to Care*. Loma Linda U.
14. "Featured Story." *Incompatible Bloodtypes*. National Kidney Foundation, 2013. www.kidney.org/transplantation/livingdonors/incompatiblebloodtype.cfm.
15. Mahdi, Batool Mutar. "A Glow of HLA Typing in Organ Transplantation." *Clinical and Translational Medicine* 2.1 (2013): 6.
16. Poulton, Kay. "Chapter 7/Transplantation Immunology: Methods and Protocols." *Methods in Molecular Biology*. By Stephen Sheldon. Vol. 333. Totowa, NJ: Humana, 2006. 157–74.
17. Erlich, H. A., G. Opelz, and J. Hansen. "HLA DNA Typing and Transplantation." *Immunity* 14.4 (2001): 347–56.
18. Schwartz, Robert S., Brian J. Nankivell, and Stephen I. Alexander. "Rejection of the Kidney Allograft." *New England Journal of Medicine* 363.15 (2010): 1451–462.
19. Eisen, Howard J. *Heart Transplantation: Graft Rejection Basics*. Johns Hopkins Advanced Studies in Medicine, May 2008.

20. Board, A.D.A.M. Editorial. *Transplant Rejection*. U.S. National Library of Medicine, 26 June 2013. www.ncbi.nlm.nih.gov/pubmedhealth/PMH0001818/.
21. Ferrara, James LM, MD, John E. Levine, MD, Pavan Reddy, MD, and Ernst Holler, MD. "Graft-versus-Host Disease." *Lancet* 373.9674 (2009): 1550–561.
22. Page, Timothy F., PhD, and Robert S. Woodward, PhD. "Cost of Lifetime Immunosuppression Coverage for Kidney Transplant Recipients." *Health Care Financing Review*30.2 (2009): 95–104.
23. Khalifian, Saami, Philip S. Brazio, Raja Mohan, Cynthia Shaffer, Gerald Brandacher, Rolf N. Barth, and Eduardo D. Rodriguez. "Facial Transplantation: The First 9 Years." *Lancet* 384.9960 (2014): 2153–63. Web.
24. Alam, Daniel, and John Chi. "Facial Transplantation." *Plastic and Reconstructive Facial Surgery* (2014): 1079–091. Web.
25. Lantieri, L., M. Hivelin, V. Audard, M. D. Benjoar, J. P. Meningaud, F. Bellivier, N. Ortonne, et al. "Feasibility, Reproducibility, Risks and Benefits of Face Transplantation: A Prospective Study of Outcomes." *American Journal of Transplantation* 11.2 (2011): 367–78. Web.
26. Shanmugarajah, Kumaran, Shehan Hettiaratchy, Alex Clarke, and Peter E. M. Butler. "Clinical Outcomes of Facial Transplantation: A Review." *International Journal of Surgery* 9.8 (2011): 600–07. Web.

CHAPTER 4
Substitutive Medicine

a. Cardiovascular and Renal Support

In all its forms, cardiovascular disease is one of the leading causes of death in the United States. Nearly one in three US adults present with one or more types of cardiovascular disease, including atherosclerosis, valvular disease, myocardial infarction, stroke, and congestive heart failure (1). According to the American Heart Association, the direct and indirect costs of major cardiovascular disease and stroke in 2011 was in excess of $200 billion in the United States and is expected to rise steadily (2). Why have cardiovascular diseases become so prevalent in the twenty-first century? Two significant reasons are changes in lifestyles and diets brought on by industrialization in the late nineteenth century. Prior to the Industrial Revolution, the majority of Americans lived in rural areas. Farming was the primary source of income and demanded extreme physical labor. Daily physical activity has been shown to be beneficial for cardiovascular health and, although the physical labor associated with farming continues to be rigorous and often grueling, the cardiovascular benefits then and now are substantial. During this time as well, food was harvested and consumed directly from local farms. Though sometimes scarce, it lacked preservatives that are commonplace today. The addition of chemical preservatives to products in the twentieth and twenty-first centuries—although necessary to prevent spoilage during processing, shipping, and storage—has been theorized to contribute to the development of several pathologies, including cardiovascular disease (3). Innovations during the Industrial Revolution also eased the burdens of daily life, and however welcomed, reduced the need for daily physical activity. This combination of a sedentary lifestyle along with increases in cigarette smoking and diets high in unsaturated fats and sugars led to increases in the prevalence of clogged blood vessels, heart attacks, and strokes. Heart disease has become fairly commonplace. The rate of heart disease increased so

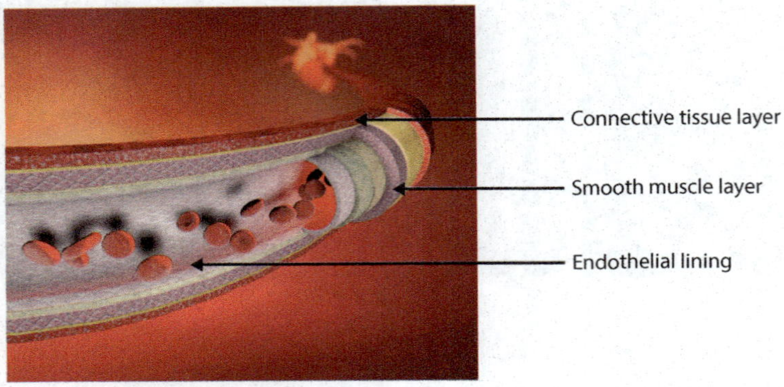

FIGURE 4.1: Coronary Artery Layers

sharply between 1940 and 1967 that the World Health Organization called it the world's most serious epidemic (4).

In 1948, a thirty-year study began in Framingham, Massachusetts. Known as the Framingham Study, the investigation involved 5,127 people aged thirty to sixty-two years who showed no signs of heart disease (5). Every two years, the participants underwent a complete physical examination. The study lasted thirty years and provided priceless profile information for predicting heart disease. Diet and exercise were found to be two of the most significant factors associated with the development of heart disease (6). Based upon these results, oral medications were developed to reduce serum cholesterol levels resulting from high-fat diets, and aspirin became a recommended prophylactic to rescue heart tissue from early ischemic events.

The most common form of heart disease in the United States is coronary artery disease (CAD), or atherosclerosis, which affects the architecture of the coronary arterial walls and ultimately vessel patency (7). Arteries are composed of three specialized tissue layers (Figure 4.1). The layer of tissue that lines the vessel lumen and is in direct contact with blood is the endothelial layer. It is composed of a single layer of specialized endothelial cells that are connected to one another by tight junctions. These cells have anticoagulant properties that prevent blood coagulation and clot formation. The peripheral adjacent layer is composed of vascular smooth muscle tissue. This muscular layer controls vasoconstriction and dilation of the vessel, thus regulating blood flow and ultimately organ perfusion. The third tissue type, which is located most distal from the arterial lumen, is the connective tissue layer that consists of the proteins elastin and collagen as well as fibroblasts. This layer provides elasticity and fixes the vessel to the surrounding tissues. Together, these specialized layers maintain laminar blood flow through the vessel, proper arterial pressures, and vessel patency.

Atherosclerosis is a disease that affects the structural architecture of these tissue layers and can begin in childhood and progress with age. Genetics may also contribute to an individual's propensity to form vascular occlusions called plaques, which are the underlying cause of coronary artery disease. Blood vessels are continually experiencing microscopic tears in the endothelium that line the lumen. Reparative processes, including clot formation and influx of inflammatory cells such as neutrophils and monocytes, ultimately resolve the damage and prepare the lumen for endothelial cell replacement and repair. As we age, our blood vessels lose the ability to readily repair vascular injuries, and inflammatory cells and blood proteins such as low-density lipoproteins (LDLs) can accumulate and remain at the sites of injury. Additionally, a certain amount of vessel wall elasticity is lost during the aging process, which exacerbates the plaque formation that leads to vessel occlusion.

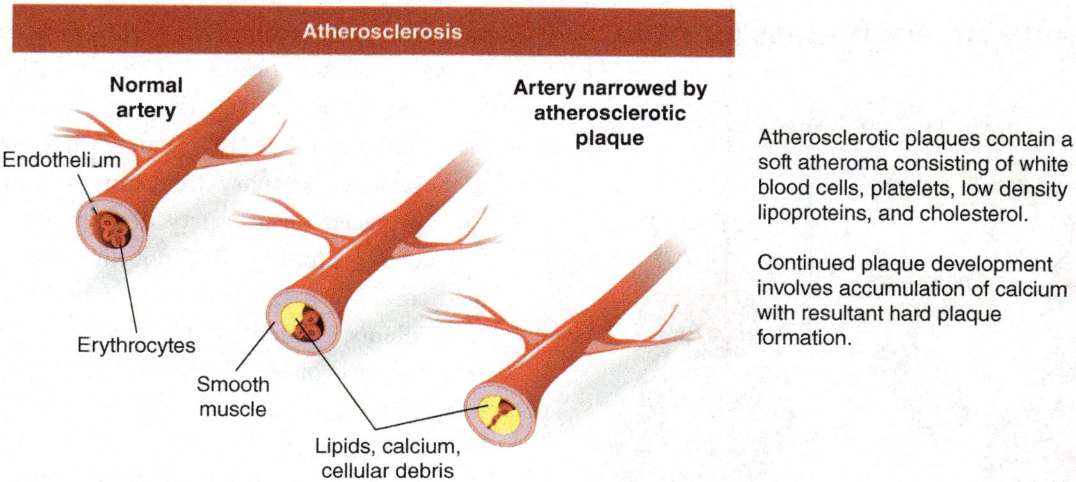

FIGURE 4.2: Atherosclerotic Plaque

A coronary blockage has three distinct components: atheroma, calcification, and cholesterol crystals (Figure 4.2). The atheroma is a nodular accumulation of a soft, flaky, yellowish material at the center of large plaques that contains a significant density of macrophages near the lumen of the artery. LDL and serum proteins collect within the atheroma over time, and underlying areas of cholesterol crystals and calcifications located at the outer base define more advanced lesions. Early lesions, consisting of soft plaques, can separate from the arterial wall and occlude the lumen downstream, either partially or totally blocking the flow of blood. Calcified plaques significantly disrupt the laminar flow of blood and can induce further clot formation and block the vessel entirely. Coronary vessels that have become completely blocked by atherosclerotic plaques lead to inadequate perfusion of distal cardiac tissue and myocardial infarction. Angina, which is defined as transient ischemia and chest pain, can result from incomplete arterial blockage that significantly limits blood flow but does not completely inhibit perfusion to distal cardiac tissue.

i. Coronary Artery Bypass Grafting

During the mid-1960s, two biomedical techniques were designed to mechanically treat atherosclerosis (clogging of the arteries). The first was coronary artery bypass grafting (CABG), and the second was balloon angioplasty, or percutaneous coronary intervention (PCI). Both techniques, still used today, were introduced clinically in 1964 and 1977, respectively (8, 9). Traditional CABG is an invasive procedure that reestablishes blood flow to cardiac tissue by grafting autologous small-diameter vessels to locations on the heart that are both proximal and distal to the atherosclerotic coronary arteries (Figure 4.3). Once anastomosed, the grafted vessels provide an alternative route for blood flow downstream to the lesion area. Prior to CABG, most patients present with symptoms such as fatigue, exercise intolerance, and angina. Diagnosis of atherosclerosis

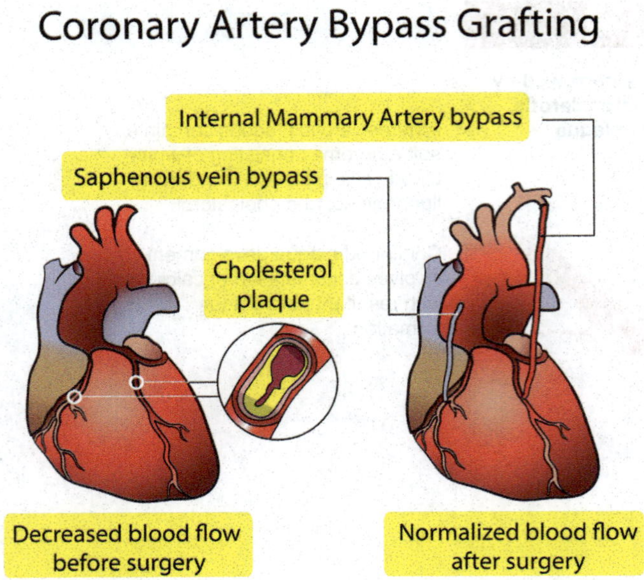

FIGURE 4.3: Coronary Artery Bypass Grafting
Copyright © by Depositphotos / Maryna Melnyk.

is based upon visualization of the coronary arteries using coronary catheterization techniques that employ contrast dyes for visual enhancement (10). Depending upon the severity of coronary occlusions, patients can be placed on a variety of therapeutic regimens, including, but not limited to, medicines such as vasodilators and anticoagulants. Patients who present with significant coronary blockages that cannot be managed pharmaceutically or by PCI (to be discussed later) undergo the CABG procedure. This procedure takes approximately six hours to complete and requires that heart function is temporarily stopped—a process called cardioplegia—and blood flow is rerouted through an extracorporeal blood membrane oxygenation (ECMO) machine that supplies oxygen and removes carbon dioxide (Figure 4.4) (11). The surgeon begins by making a 30-cm midline incision in the chest and performs a sternotomy, ultimately separating the rib cage in order to access the heart and the afflicted vessels. Replacement vessels are then harvested from either the internal mammary arteries or saphenous veins, or both, and sutured in place as conduits bypassing the coronary vessel occlusion. Once the grafted vessels are secure, the heart is stimulated to begin beating and the patient is removed from ECMO, although percutaneous tubes remain in place for several days in order to drain excess fluid from the thoracic cavity. Time on ECMO varies patient to patient and depends upon native heart function once the heart has been induced to beat on its own. The cost of the CABG procedure alone is approximately \$35,000–\$40,000 (12). Although patients must remain in the hospital for seven to ten days postoperatively and the rehabilitative period

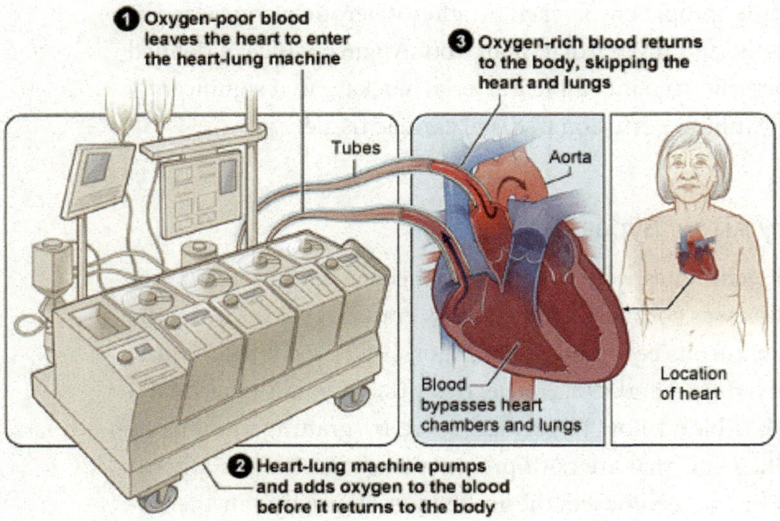

FIGURE 4.4: Heart–Lung Machine
NHLBI / NIH / Copyright in the Public Domain.

TABLE 4.1: Comparison of CABG, MIDCAB, and OPCAB (Original Table)

	CABG	MIDCAB	OPCAB
Length of incision	30 cm	10 cm	20–30 cm
Maximum number of arteries that can be grafted	5	1–2	4
Operation on a beating heart?	No	Yes	Yes
ECMO required?	Yes	No	No
Operative mortality	2.9%	1.5%	0.8%

can take from ten weeks to six months, up to five arteries can be bypassed, and grafts can remain patent for up to twelve years postsurgery. Primary postsurgical complications include infections, bleeding, and thrombosis. Immune function is temporarily impaired as a result of significant surgical trauma, and patients may be more susceptible to secondary infections during recovery. Patients may also experience cognitive deficits such as post-perfusion syndrome that may be associated with extracorporeal blood perfusion and exposure to the oxygenation membrane. The mortality rate is approximately 2.9 percent due to the extent of preexisting cardiac ischemia, invasiveness of the procedure, and time on ECMO (12).

Alternatives to traditional CABG have been developed that are less invasive and do not require extracorporeal blood oxygenation (Table 4.1). The first is off-pump coronary artery bypass (OPCAB) (13). The significant difference between traditional CABG and OPCAB is that OPCAB can be performed on a beating heart, thus removing ECMO as a necessary part of the procedure. The surgeon makes a midline incision that can vary in length from 20–30 cm and a medial sternotomy is performed. Adenosine and esmolol are used to slow the heart rate. To allow for access to the entire heart, there must be a sufficient amount of cardiac displacement, which is accomplished by deep pericardial sutures and the use of specialized instruments to prop the heart in a position that will allow the surgeon to access occluded arteries. Along with sponges, some surgeons use slings to prop the heart in the necessary positions and then employ a stabilizer to focus on a particular occluded artery. Vascular grafts are harvested from the internal mammary arteries and saphenous veins following a protocol similar to that for CABG. With the heart still beating, there is greater difficulty in performing a bypass on the posterior and lateral walls of the heart; therefore, a maximum of four coronary arteries can be bypassed during OPCAB as opposed to five arteries during CABG. The second alternative to CABG is minimally invasive direct coronary artery bypass (MIDCAB) (14). Grafting is performed on a beating heart much like OPCAB; however, the incision is significantly smaller, just 10–12 cm. Grafts must be sutured under direct vision, and the coronary artery to be bypassed must lie directly beneath the incision line. This further limits the number of possible vessels that can be grafted to two. Both OPCAB and MIDCAB are considerably less expensive than traditional CABG, costing 25 percent and 40 percent less, respectively. Additionally, both procedures avoid the risks associated with ECMO because grafting is completed on a beating heart and reanimation is not necessary. Time in surgery, in-hospital stays, recovery times, and mortalities are all reduced by approximately 50 percent as compared to CABG. Although OPCAB and MIDCAB present two viable alternatives to traditional CABG, many patients and their physicians choose the traditional approach due to the specialized compulsory surgeon training and the limited number of vessels that can be grafted during the procedure.

Bypass grafting, although effective at establishing adequate perfusion of cardiac tissue, is not a cure for cardiovascular disease. Graft patency is variable, and viability of grafted vessels averages approximately ten to twelve years (15). Further bypass grafting is often not a viable option, and patients must then resort to implantation of cardiac perfusion support devices, such as ventricular assist devices or heart transplant surgery.

Ten years following the introduction of CABG, a catheter-based procedure—percutaneous coronary intervention (PCI), or angioplasty—was developed and used clinically (17, 18). The goal of PCI is to unblock partially occluded coronary vessels without having to perform open heart surgery. By doing so, patients preserve their native coronary vessels and avoid the risks associated with CABG, such as post-perfusion syndrome. This procedure benefits all patients, but especially younger cardiac patients who may ultimately require bypass grafting but can delay such invasive procedures for several years.

ii. Percutaneous Coronary Intervention

Prior to the first series of cardiac catheterizations, in 1958, Dr. F. Mason Sones of the Cleveland Clinic discovered that contrast dye could be injected into the heart without harm to a patient (16). This discovery led to the angiogram, a test used to gather images of the heart's blood vessels. Angiograms allow for visualization of blood flow through the affected coronary arteries and are therefore essential for proper diagnosis and treatment of the atherosclerotic vessels. Shortly following this discovery, Dr. Charles Dotter in 1964 in Portland, Oregon, used catheters to open blockages in peripheral arteries; and in 1977, Dr. Andreas Gruentzig performed the first balloon angioplasty procedure in a human patient in Switzerland (17, 18). The first documented human catheterization was performed by Dr. Werner Forssmann on himself in 1929 (19). Fifty years passed before cardiac catheterization became clinically approved in the United States in the mid-1970s.

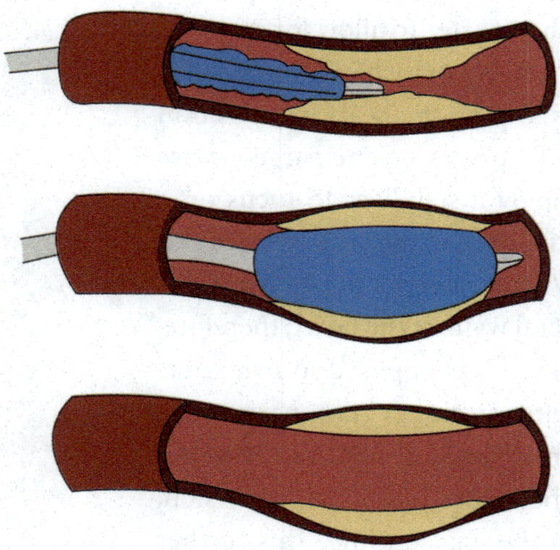

FIGURE 4.5: Percutaneous Coronary Intervention
Copyright © by Depositphotos / turhanerbas.

PCI is now the most commonly used cardiovascular intervention, and approximately 1.5 million patients receive treatment each year in the United States (20). PCI is performed by a cardiologist on an outpatient basis (Figure 4.5).

The cardiologist guides a catheter through the arterial vasculature to the location of the coronary plaque using a guide wire that has been previously inserted through the femoral artery. A balloon, composed of elastic polymer material, is attached to the end of the catheter and is inflated once it is positioned within the partially occluded region of the vessel. The balloon is inflated and deflated several times in order to mechanically press the plaque material against the luminal wall of the vessel. Once patency is established, as determined by fluoroscopy or intravascular ultrasound, the balloon and catheter are removed.

PCI has been a successful first-line treatment for patients experiencing coronary blockages due to atherosclerosis. This technique, however, does carry associated risks that negatively affect the treated vasculature. Acute vessel closure can occur within minutes after the procedure as a result of dissection and elastic recoil of the artery, leading to acute myocardial infarction (AMI) and emergency CABG. Additionally, denudation of the endothelial wall can lead to platelet aggregation and thrombosis, further exacerbating acute vessel closure. Patients are also susceptible to later-phase occlusions, and approximately 30 percent to 40 percent of patients who are treated with coronary angioplasty present with additional blockages in the same locations within one year of the procedure (21). These late secondary blockages are not composed of plaque material but are the results of mechanical and biological processes. These processes collectively are referred to as *postangioplasty restenosis* and include the mechanisms of neointimal hyperplasia and elastic recoil (Figure 4.6) (22).

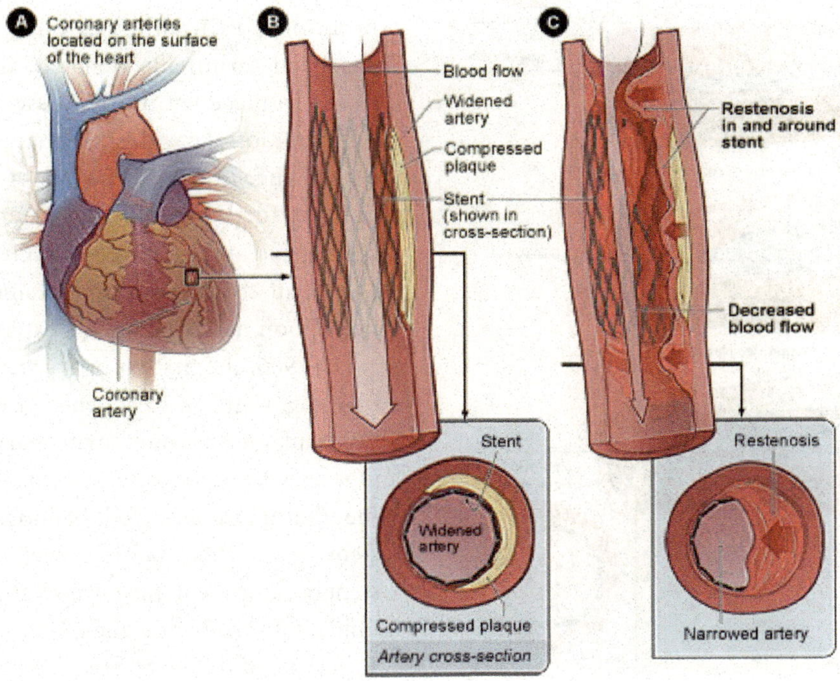

FIGURE 4.6: Post-PCI Restenosis
NHLBI / NIH / Copyright in the Public Domain.

When a coronary balloon is inflated in the location of an atherosclerotic plaque, biological materials comprising the plaque, such as calcified fatty deposits, are mechanically pressed against the underlying endothelial wall. This procedure, although effective at reestablishing transient luminal patency, results in microscopic endothelial tears and early thrombosis. The cells of the underlying exposed smooth muscle layer become activated when in contact with the circulatory environment, and the affected smooth muscle cells begin to proliferate. Coinciding with smooth muscle cell proliferation is production of extracellular matrix and deposition into the lumen of the vessel. Together with the mechanical effects of elastic recoil, the lumen of the treated coronary artery becomes occluded, and blood flow is again compromised. In order to prevent this fairly common occurrence, bare metal stents and drug-eluting stents have been designed and developed through several generations. Approximately 70 percent of patients who receive PCI are stented as well (23).

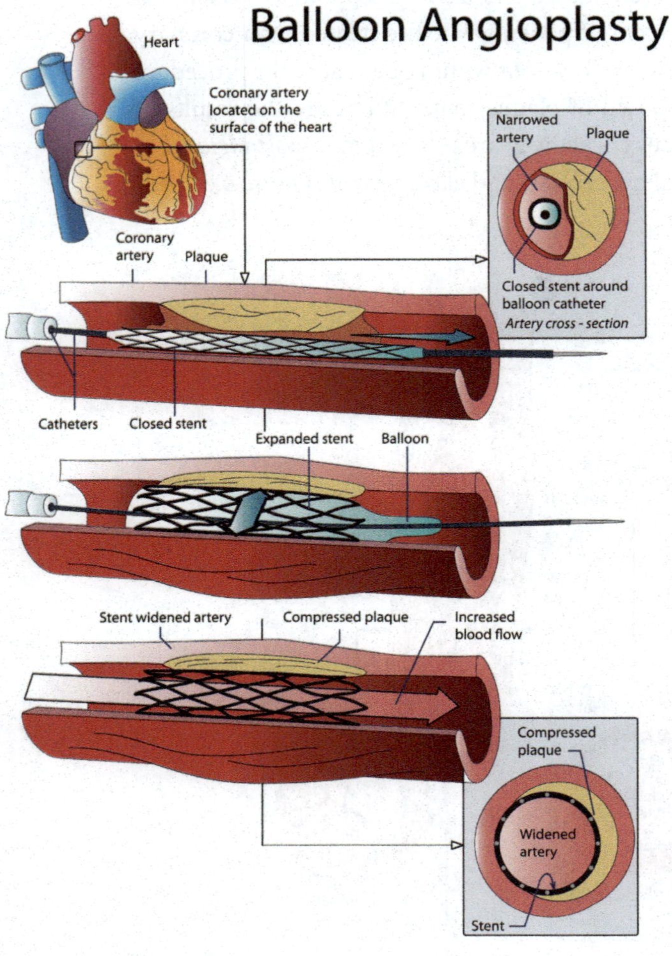

FIGURE 4.7: PCI with Stent

Copyright © by Depositphotos / Maryna Melnyk.

iii. Coronary Stents

A coronary stent is effectively a corrugated circular metal tube that surrounds a PCI balloon at the tip of the catheter that is designed to prevent restenosis following PCI (Figure 4.7). Catheterization and insertion of the stent follows a similar protocol as traditional PCI, although once the balloon has been terminally deflated, the stent remains in place within the vessel. Permanent implantation provides mechanical support within the lumen of the vessel in order to prevent further occlusions. Stent walls are corrugated and can be constructed with variable strut thicknesses and designs and may be balloon deployed or self-expandable (24).

In 1986, the first coronary stent, a self-expanding, stainless-steel, wire-mesh construct, was implanted in a human coronary artery (25). One year later, Johnson & Johnson developed the first FDA-approved balloon-expandable stent (26). This stainless-steel, slotted-tube stent became the mainstay for bare-metal stent implantations during the 1990s. Many other variations of this stent design were developed, and all reduced the incidence of elastic recoil and early restenosis. These early stents were associated with subacute in-stent thrombosis due to their high metallic densities and bulk, although several clinical trials conducted during this time period did demonstrate clinical superiority of PCI with stent versus PCI alone (27, 28). Currently, implanted bare-metal stents are composed of metals such as cobalt-chromium alloy and nitinol instead of the original stainless steel (29).

Long-term studies of patients implanted with bare-metal stents followed on the original studies that demonstrated physiological success of PCI with stenting as compared to PCI alone (30, 31). These studies, however, showed a 20 percent to 30 percent incidence of late-stage in-stent restenosis at time points up to a year postimplantation (31). In-stent restenosis at these later points results from vascular smooth muscle activation, proliferation, and extracellular matrix production (32). As with PCI, mechanical shearing of the endothelial lining during stent deployment results in exposure of the underlying smooth muscle layer. When exposed to the circulatory environment, vascular smooth muscles are induced to proliferate and deposit extracellular matrix into the lumen of the vessel. In order to address this issue, various compounds were investigated as stent coatings to target luminal inflammation, platelet activation, thrombosis, and vascular smooth-muscle-cell proliferation. Two compounds, siromilus and paclitaxel, were shown to significantly reduce the incidence of late-stage in-stent restenosis when coated onto bare-metal stents (Figure 4.8) (33, 34). Both agents interfere with cell division, albeit through different cellular pathways. Siromilus, an immunosuppressive compound derived from the Rapa Nui fungus found on Easter Island, blocks transition through the cell cycle; whereas paclitaxel, an anticancer drug derived from *Taxus brevifolia*, the Pacific yew tree, interferes with microtubule formation during the cell cycle (34). In the first generation of drug-eluting stents (DES), the drugs were incorporated into permanent polymer coatings of polyethylene-co-vinyl acetate and poly-*n*-butyl methacrylate (35). These first-generation stents were implanted in 1999 and became clinically available in 2002 (36, 37). Second-generation stents employ more biocompatible polymer coatings such as poly-vinylidene fluoride and hexafluoropropylene (36). In 2006, safety issues emerged regarding the presentation of late-stage stent thrombosis potentially resulting from inhibition of re-endothelialization and delayed immune reactivity to the polymer substrate (38). Although these concerns were brought to the attention of the media and the FDA, further investigations revealed that the benefits of DES outweigh the low incidence of late-stage thrombosis and therefore DES should continue to be clinically available (39). Further development efforts have focused on designing bioresorbable drug-eluting stents composed of polymers such as poly-L-lactic acid and poly-DL-lactic acid. Additionally, drugs that initiate accelerated healing and

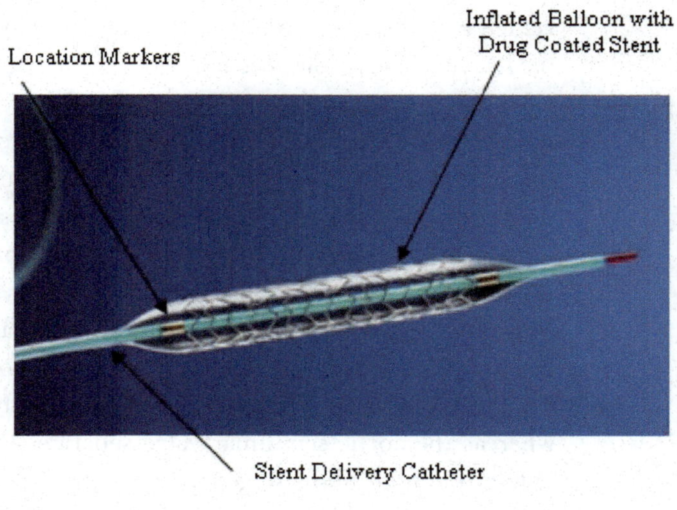

FIGURE 4.8: Drug-Eluting Stent
FDA / Copyright in the Public Domain.

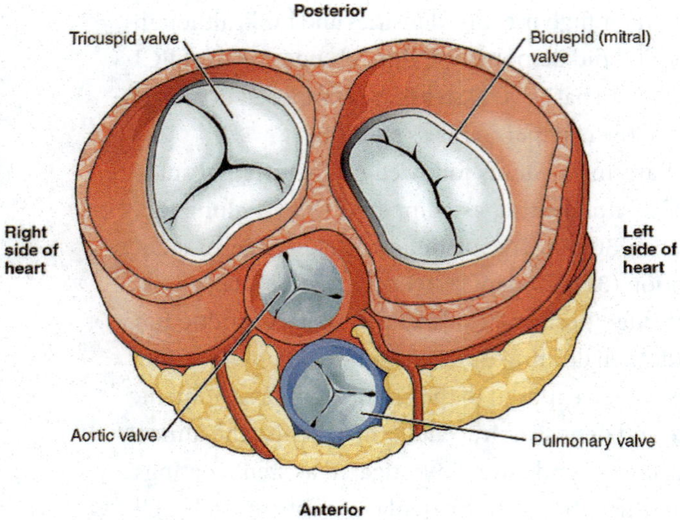

FIGURE 4.9: Cardiac Valves
Copyright © 2013 by Rice University, (CC BY 3.0).

promote endothelialization are being investigated as novel drug-stent combinations.

iv. Cardiac Valves

As blood is pumped through the human heart, four cardiac valves function to ensure unidirectional blood flow. The cardiac valves—the tricuspid, the pulmonary semilunar, the bicuspid, and the aortic semilunar—are passive structures composed of fibrous connective tissue and low densities of fibroblasts that are oxygenated by the blood circulating in their path (Figure 4.9). All four valves perform their functions passively as a result of pressure differentials between the chambers and associated vasculature. The tricuspid valve is located between the right atria and the right ventricle. It is composed of three leaflets that are tethered to the wall of the right ventricle by Purkinje fibers. During the resting phase (diastole) of the cardiac cycle, the right ventricle is relaxed, pressure is low, and blood flows passively through the tricuspid valve from an area of higher pressure in the right atrium to an area of lower pressure in the right ventricle. As the right ventricle fills and begins to contract (systole), the intraventricular pressure rises and surpasses that of the right atrium and the tricuspid valve closes and is secured in place by the tethering Purkinje fibers. Retrograde blood flow is prevented by the secure closure of the valve. The bicuspid valve that separates the left atrium from the left ventricle is composed of two leaflets that are tethered in a similar fashion as the tricuspid valve and functions in a similar manner. The pressures on the left side of the heart, however, are significantly higher than those on the right side due to the greater resistance of the peripheral circulation than the pulmonary circulation. This places the bicuspid valve at greater risk than the tricuspid valve for damage and failure. The semilunar valves, although composed of similar biological materials as the bicuspid and tricuspid valves, have different mechanical structures. Both valves have three leaflets that are shaped as shallow cups that are attached at their bases to a ring of tissue, the annulus, which connects the valves to the heart muscle. The edges of the leaflets meet in the center of the valve and open and close in response to pressure differences on either side of the valve. The pulmonary semilunar valve separates the right ventricle from the lungs, whereas the aortic semilunar valve separates the left ventricle from the systemic circulation. Due to the high pressure generated by the left ventricle and high resistance of the peripheral vasculature, the aortic valve experiences greater wear and tear than the other three valves and is most often the valve that requires replacement. Damage to this valve can be catastrophic, because it is critical in supporting left ventricular pressure during isovolumetric contraction and is the only gateway for blood out of the heart into the systemic circulation.

Cardiac valve disease can be categorized as either valvular stenosis or valvular insufficiency (40). As a result of high left ventricular pressure and peripheral resistance, the aortic semilunar valve is most often the valve affected. Valvular stenosis can be described as stiffening of the valve leaflets and can result from calcium deposition or secondary infection, which causes the subsequent inability of the valve to open and close properly (Figure 4.10). Blood pumped through the valve into the aorta by the left ventricle is ejected through the narrowed opening of the valve with reduced ejection volume during each cardiac cycle. Hemolysis (red blood cell rupture) and thrombosis can occur due to resultant turbulence in blood flow. Insufficiency is a second type of cardiac valve disease caused by conditions such as congenital malformation, infection, and medications or is idiopathic in nature. Valves that are insufficient lack structural integrity and are unable to close properly, thus allowing regurgitation of blood into the adjacent cardiac pumping chamber (Figure 4.11). Both valvular disorders eventually lead to decreased stroke volumes (amount of blood ejected from the left ventricle during each systolic contraction) and increased heart size (congestive heart failure) as the cardiac muscle cells attempt to compensate for the reduced stroke volume. Patients experiencing either pathology are typically treated with first-line medications that are designed to strengthen the force of cardiac contraction, along with anticoagulants to prevent blood clot formation. As the disease state progresses, cardiac valve replacement is recommended and often necessary. Cardiac valves can be replaced by either mechanical valves or tissue-based valves.

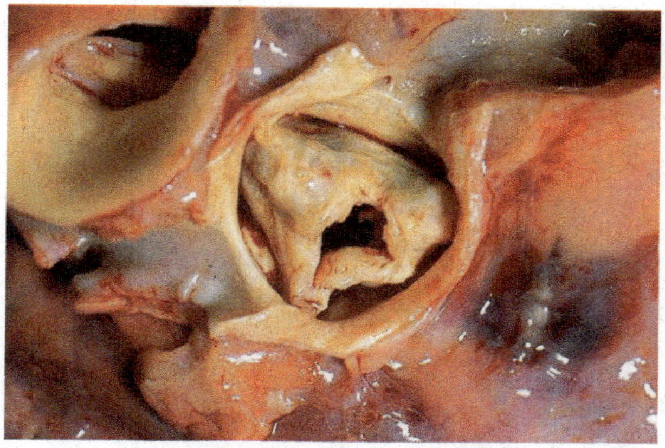

FIGURE 4.10: Aortic Valve Stenosis
Edwin P. Ewing, Jr. / CDC / Copyright in the Public Domain

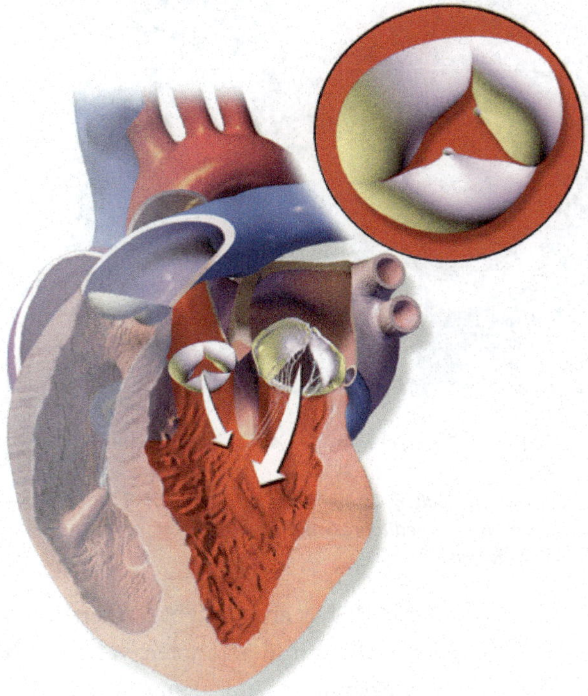

FIGURE 4.11: Valvular Insufficiency
Copyright © 2013 by BruceBlaus / Wikimedia Commons, (CC BY 3.0) at http://commons.wikimedia.org/wiki/File:Blausen_0039_AorticRegurgitation.png.

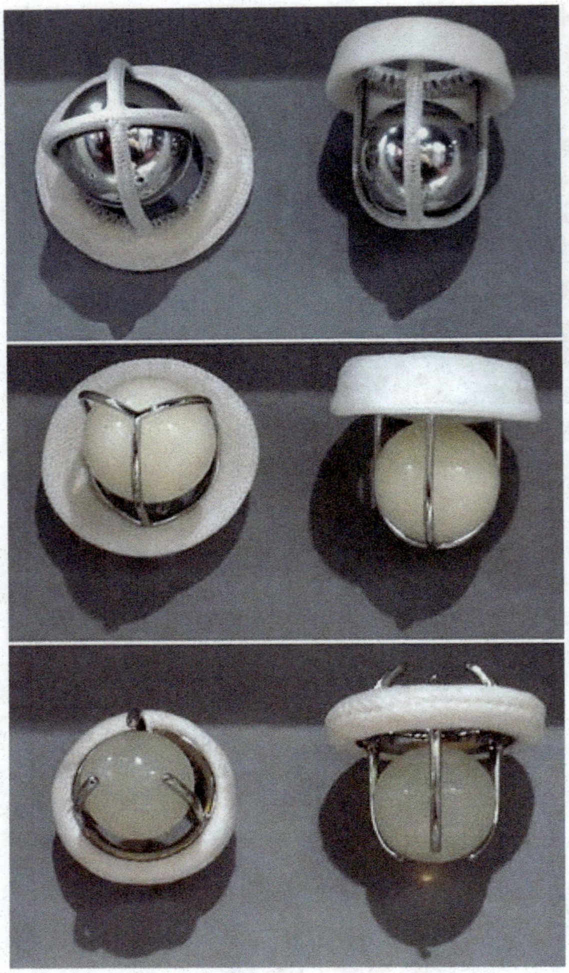

FIGURE 4.12: Mechanical Cardiac Valve: Caged-Ball Valve

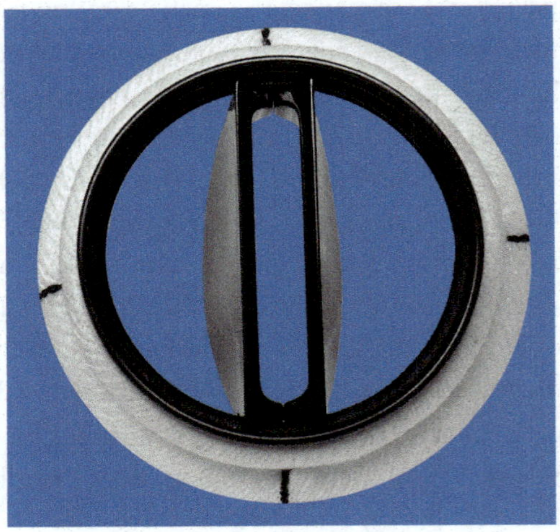

FIGURE 4.13: Tissue-Based Prosthetic Valve and Mechanical Bi-Leaflet Valve

Albert Starr and Dwight Harken performed the first implantations of mechanical caged-ball valves both at the mitral and aortic valve positions during the early 1960s (41). Design criteria included size, durability, chemical inertness, and hemocompatibility (the absence of substantial adverse events in the vascular system). All criteria allow for long-term function and were immediately realized, and several variations of mechanical cardiac valve replacements were developed. The Starr-Edwards caged-ball valve was followed by the development of leaflet-based valves, such as St. Jude's bileaflet mechanical valve and Medtronic's single-leaflet mechanical valve (Figures 4.12, 4.13) (42). Although varied in structural design, each type of prosthetic valve contains a combination of alloyed metal, such as stainless steel and cobalt-chromium; plastics, such as polyethylene terephthalate (Dacron)

TABLE 4.2: Mechanical vs. Tissue-Based Prosthetic Cardiac Valves (original table)

Valve Type	Pros	Cons
Mechanical	Highly durable Can be implanted in young patients	Increased risk of thrombosis Lifetime anticoagulant therapy required
Tissue based	Anticoagulants and immunosuppressants not required	Limited durability of approximately 12 years Potential calcification Potential for infection

and silastic elastomer; and coatings, including pyrolytic carbon. The alloy metal frame provides the support for either the caged-ball or synthetic leaflets of the valve, and Dacron provides a hemocompatible surface for suturing the base of the valve onto native tissue. The ball of the caged-ball valve can be composed of either silastic elastomer or metal alloy, and the leaflets of both the uni- and bileaflet valves consist of pyrolytic carbon. Mechanical valves typically last more than twenty years and provide successful replacements for diseased native valves. The major contraindication for use is mandatory anticoagulant therapy. Patients prone to bleeding and who are averse to anticoagulant therapies cannot be implanted with this type of replacement valve (Table 4.2).

Patients seeking an alternative to mechanical valves may be qualified to receive tissue-based valve implants (43). Prosthetic tissue valves combine both synthetic and natural components to provide patients with valve replacements that do not require anticoagulant therapy (Figure 4.14) (44). Stented valves are constructed with metal alloy supports that provide a framework for attachment of the tissue valve leaflets and Dacron sewing ring. Stented valves include the Hancock and Carpentier-Edwards porcine leaflet and bovine pericardial leaflet valves, respectively. Prior to manufacture, the xenogeneic tissue is treated with formaldehyde in order to cross-link native proteins located on the valve surfaces that may be antigenic to patients. Stentless valves are also available that employ similar components as stented valves, although they are not fabricated with alloy-based strutted frames and therefore typically do not possess the same level of mechanical strength as stented valves. Although tissue-based valves are good options for patients who cannot tolerate anticoagulant therapy, these valves are associated with higher percentages of infection, have reduced functional time spans of ten to fifteen years, and are prone to calcification, which can lead to valve stenosis of the replacement (Table 4.2).

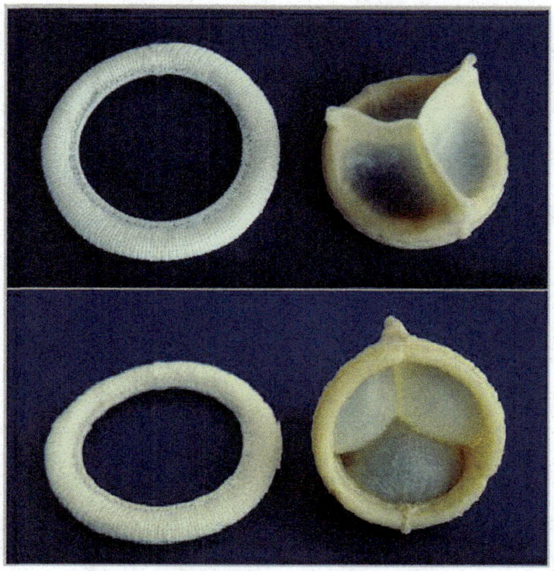

FIGURE 4.14: Prosthetic Tissue Valve

v. Left Ventricular Assist Devices

As seen in previous sections, several cardiac disorders can be treated by repairing or replacing the components that have become diseased. In attempts to compensate for overall decreases in cardiac function, cardiac muscle cells enlarge, most prominently in the left ventricle, resulting in congestive heart failure. The enlargement of cardiac muscle cells can transiently compensate for decreases in contractility by increasing the overall force of contraction and increasing the volume of blood ejected during each cardiac cycle. As cardiac cells continue to hypertrophy (enlarge), the contractile mechanisms resulting from actin and myosin cross-linking become less effective and eventually fail. This can be clearly represented on a Frank-Starling curve (Figure 4.15) (45). Patients are eventually treated with ionotropic medications to further increase the force of contraction and stroke volume. Ultimately, the pumping function of the heart—particularly the left ventricle—requires replacement, and this can be accomplished by a ventricular assist device (VAD) (Figure 4.16). VADs are mechanical pumps designed to offload the ventricles and deliver blood to the body (46). Three main conditions warrant the use of VADs (Table 4.3). The first indication for use is as a "bridge to repair." Patients who require temporary offloading of their ventricles suffer from reversible conditions such as viral infections that temporarily disrupt cardiac function. If the native ventricle is allowed to rest while the VAD assumes its function, biochemical and mechanical processes are able to repair. Following a prescribed amount of time, the native ventricle resumes full function. The second purpose for ventricular support

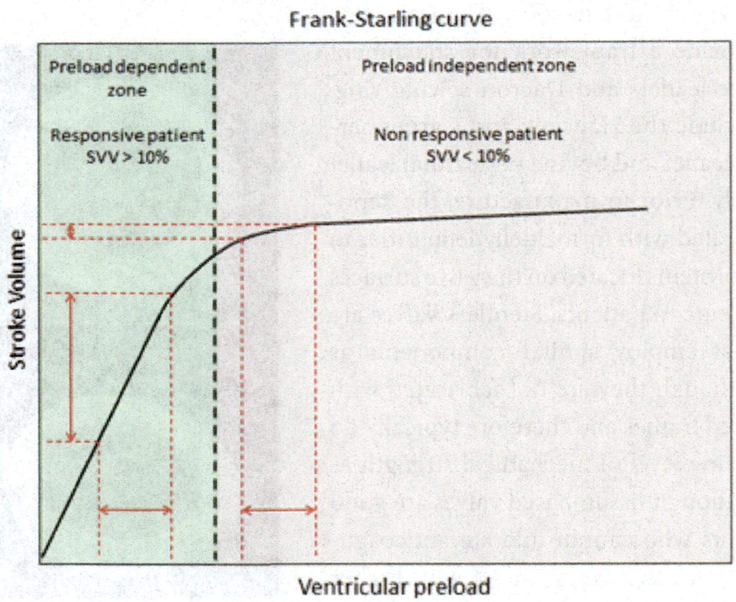

FIGURE 4.15: Frank Starling Curve of Left Ventricular Heart Function
Copyright © 2015 by Patricia Pineda Vidal, (CC BY-SA 4.0) at https://commons.wikimedia.org/wiki/File:Frank_Starling%E2%80%99s_curve.png.

is "bridge to transplant." The eventual option for patients suffering from congestive heart failure is heart transplantation. Many of these patients have additional organ dysfunction due to their preexisting conditions and require adequate blood perfusion while awaiting lifesaving transplant. Support by a ventricular assist device provides sufficient pumping with resultant perfusion of peripheral organs. This can optimize patient health for successful transplantation. Destination therapy, terminal implantation of a ventricular support device, is the ultimate goal of ventricular support and total artificial heart programs. The principle of destination therapy is to replace native cardiac function on a permanent basis by delivering optimal circulatory support using a completely implantable mechanical replacement, thus negating the need for transplantation. The significant difference between a VAD and a total artificial heart is the necessity to remove the patient's natural heart when implanting the total artificial heart. Implantation of a VAD does not require removal of the patient's heart (47). Today, most VADs are used as bridges to transplant and not destination therapies.

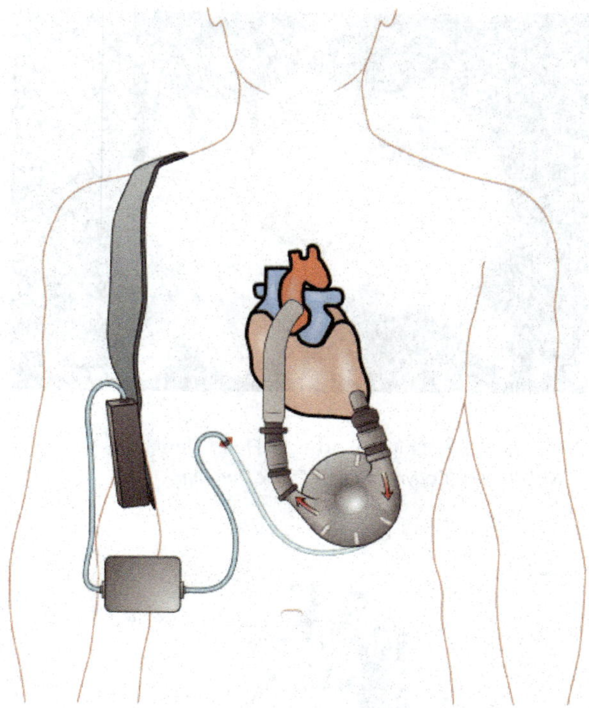

FIGURE 4.16: Ventricular Assist Device

In 1982, Dr. Barney Clarke was the first patient to be implanted with a totally implantable artificial heart designed by Dr. Robert Jarvik, and the first pneumatically powered VAD for clinical use was approved by the FDA in 1994 (48). The large external console controlled both power and function, and the pulsatile pumping mechanism was encased in a fist-size polycarbonate

TABLE 4.3: Clinical Applications for Ventricular Assist Devices (original table)

Application	Description
Bridge to recovery	Temporary cardiac support that allows the native heart to rest and ultimately heal. VAD is removed when the native heart can function on its own.
Bridge to transplant	Temporary cardiac support for patients awaiting heart transplant surgery. VAD removed as patient is being connected to extracorporeal support and donor heart is transplanted.
Destination therapy	Permanent cardiac support for patients who are ineligible for transplant and whose hearts will not recover on their own. Native heart is removed and replaced with permanent VAD.

FIGURE 4.17: Original Artificial Heart (Jarvik-7)
NHLBI / NIH / Copyright in the Public Domain.

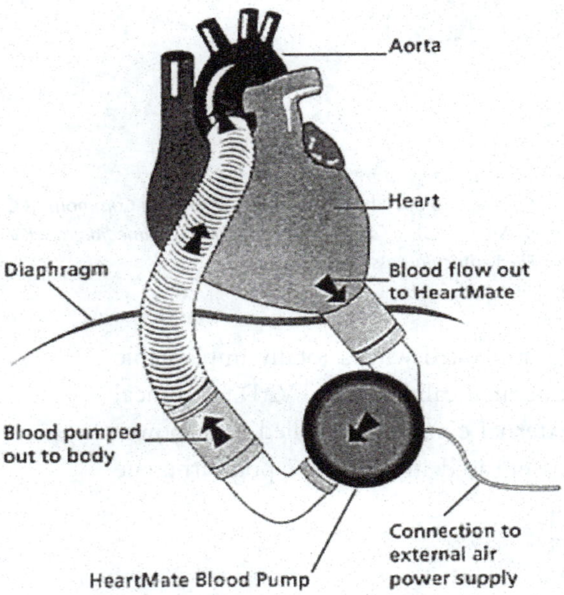

FIGURE 4.18: Components of a Left Ventricular Assist Device
Steven M. Gordon / CDC / Copyright in the Public Domain.

implantable chamber (Figure 4.17). The internal bladder that received and then pumped blood was fabricated from polyurethane, the flexible polymer. In 1998, an electrically powered VAD was approved by the FDA that was significantly more portable (49). Over the past thirty years, substantial improvements have been made to optimize the design and function of VADs and totally implantable artificial hearts. Design modifications have been made to improve hemocompatibility by using more biocompatible materials and by converting from pulsatile flow to axial flow systems that are significantly smaller in size (50). Devices with both flow mechanisms are still available for use.

Although variable in design, ventricular assist devices share several basic components (Figure 4.18). Blood is siphoned from the heart through a flexible Dacron vascular graft connected to the apex of the left ventricle. Blood flows from the left ventricle, through the Dacron graft, and into the pumping chamber of the VAD. Blood is then pumped back to the patient via an outflow Dacron graft that is anastomosed to the patient's arterial system. Cardiac output can be regulated to provide patients with up to 12 liters/minute of blood flow. Early pulsatile designs have been partially replaced by axial flow devices that can be fabricated on a much smaller scale with reduced risk of clot formation. VADs that are fully implanted are typically used for patients who are mobile and able to return home. A percutaneous driver line connects the internal device to the battery pack and controller that are secured to a belt or back strap on the patient. The external lithium ion battery pack can be charged on a wall unit when necessary. VADs that are not fully implanted require that the pumping chambers as well as the driver line, controller, and battery pack are located extracorporeally and only the Dacron vascular grafts are implanted. Patients are immobile and usually are awaiting transplant. The AbioCor totally implantable artificial heart, developed by AbioMed, Inc., has avoided having to implant percutaneous driver lines by powering the device using a transcutaneous energy transmission system (Figure 4.19) (51). This particular design modification is critical for avoiding one of the most common adverse events associated with VADs and the total artificial heart: infection.

FIGURE 4.19: AbioMed Total Artificial Heart

Infection, thrombosis, and subsequent stroke are the major adverse events associated with VAD implantation. The percutaneous driver line breaches the protective skin barrier and provides a port of entry for bacterial contamination. Immediate infections, initiated during surgical implantation, and those that present at later points can result from inadvertent relocation of bacteria into the surgical site and colonization along device components. Very rarely can bacterial contamination of an implanted device be controlled by exogenous administration of antibiotics. Most often, the device must be removed and the implantation site debrided in order to control further colonization. A second adverse event, perhaps more prevalent than infection, is thrombosis. Although patients are maintained on anticoagulants during and after surgery, blood clots can develop on any of the internal, blood-contacting components. Clots that form typically break off from their original contact point and travel along with the blood, eventually becoming trapped in the patient's vasculature. Clots that travel through the carotid arteries to the brain are capable of causing stroke and permanent neurological damage. Innovations in device design have led to the development of smaller VADs with few blood-contacting components that are suitable for long-term implantation with lower risks of infection and thromboembolism.

b. Dialysis

Approximately seventy thousand patients per year in the United States are on the organ transplant list awaiting donor kidneys, while the average number of kidney transplants per year in the United States is eighteen thousand, including transplants from living donors (52). This disparity between the supply and demand has warranted the use of dialysis for extracorporeal kidney support. Dialysis—filtration of the blood to remove excess fluid, uremic toxins, and salts—is reserved for patients who present with end-stage kidney disease (ESRD). Although the kidneys perform additional functions, such as acid-base balance and blood pressure control, dialysis is designed to specifically regulate patient fluid volumes and to detoxify the blood. The kidneys are two fist-sized organs that each contains 1 million filtration units, nephrons, connected directly to individual arterial blood supplies (Figure 4.20) (53). During a twenty-four-hour period, each kidney filters an average of 180 liters of water, 600 grams of sodium, 180 grams of glucose, and 50 grams of urea. Much of the water, sodium, and glucose are reabsorbed back into the body, while approximately

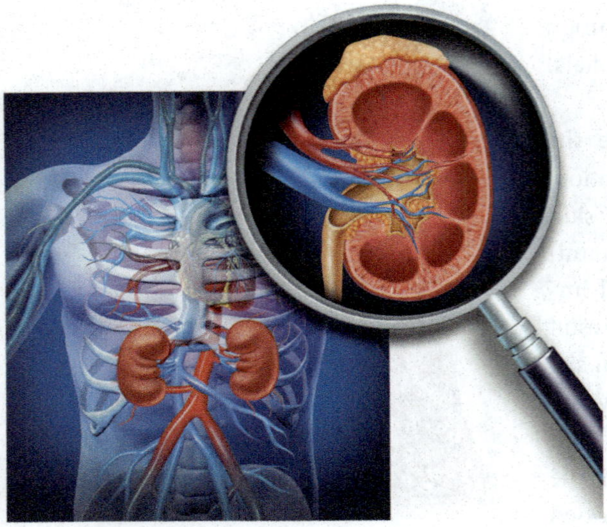

FIGURE 4.20: The Kidney

50 percent of the urea is excreted (Table 4.4). Kidney function, specifically urine and plasma levels of urea and creatinine, can be determined by using diagnostic tests such as clearance and glomerular filtration rate (Figure 4.21). Clearance is the ratio of the concentration of urea or creatinine in the blood to that in the urine over a specified period of time. Further determination of kidney function can be assessed by glomerular filtration rate. In this diagnostic test, a starchlike compound, inulin, is injected into the patient and the concentration in the urine as compared to that in the plasma is quantified over a designated time interval. The glomerular filtration rate (GFR) for a healthy individual is 125 mL/minute. Patients first become symptomatic when clearance falls to about 10 percent of normal levels and critically ill at 5 percent of normal levels.

TABLE 4.4: Kidney Filtration and Reabsorption of Key Substances (original table)

Key Substance	Amount Filtered in 24 Hours	Percentage Reabsorbed in 24 Hours	Amount Excreted in 24 Hours
Water	180 L	99%	1.8 L
Urea	54 g	44%	30 g
Sodium	630 g	99.5%	3 g
Glucose	180 g	100%	0

The rate of excretion of the urea and creatinine are used to determine renal clearance and general kidney function.

Clearance = Urea or creatinine concentration in urine × Volume/concentration in plasma
**over a known time interval*

The filtration of exogenously administered inulin is used to determine glomerular filtration capacity.

Glomerular filtration rate = Urine concentration (inulin) × Urine flow/plasma concentration (inulin)

FIGURE 4.21: Clearance and Glomerular Filtration Rate (GFR) Equations (Original Figure)

Renal dysfunction is categorized as either acute kidney failure or chronic kidney failure. Both types of failure can occur due to prerenal causes that result in significantly reduced blood flow to the kidney as well as direct kidney damage. Hypovolemia, low blood and tissue volumes, is the root cause of prerenal failure and can result from dehydration, medication-induced fluid loss, and arterial and venous obstruction of blood flow to the kidneys. Direct kidney damage is also a significant contributor to kidney disease. Trauma, medication-induced cellular damage, infection, and high blood pressure can all result in direct damage and long-term failure. Once diagnosed, patients have several treatment options, including kidney transplant, hemodialysis, and peritoneal dialysis.

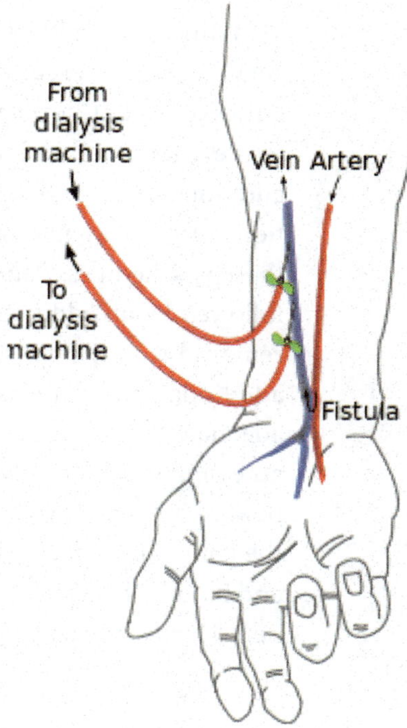

FIGURE 4.22: Arteriovenous Fistula for Dialysis Access

Copyright © 2008 by Kbik / Wikimedia Commons, (CC BY-SA 3.0) at http://commons.wikimedia.org/wiki/File:Radiocephalic_fistula.svg.

As stated previously, organ transplantation is not always an available or appropriate option. Dialysis, filtration of the blood by synthetic means, remains the most common treatment modality prescribed for end-stage renal patients. There are over five hundred thousand ESRD patients in the United States, and approximately 77 percent are maintained on dialysis therapy (54). In the United States, dialysis centers are for-profit companies, and a year of dialysis costs over $70,000 for Medicare patients and over $180,000 for those covered by employer group health plans (55). Coverage, however, is provided by the government regardless of age and need.

Patients who are prescribed hemodialysis, direct filtration of the blood against a selective membrane, typically undergo therapy three days per week, although this may vary according to the patient's needs. Prior to commencement of treatment, access to the vasculature must be established. Vascular access is required for arterial and venous connections. This can be established in two ways (56). An arteriovenous fistula can be surgically created by connecting an artery to a vein in the patient's forearm (Figure 4.22). The anastomosis must be made at least a year prior to the beginning of dialysis treatments in order to allow remodeling of the vein. Blood pressure from the artery causes dynamic changes in the connected vein, which results in thickening of the smooth muscle layer of the vascular wall and development of a more arterial-like structure. This procedure provides more access sites for the dialysis needles and lessens the potential for vascular collapse and blood clot formation. Implantation of a synthetic vascular graft that connects a vein and artery is a second method for establishing repetitive vascular entry (Figure 4.23).

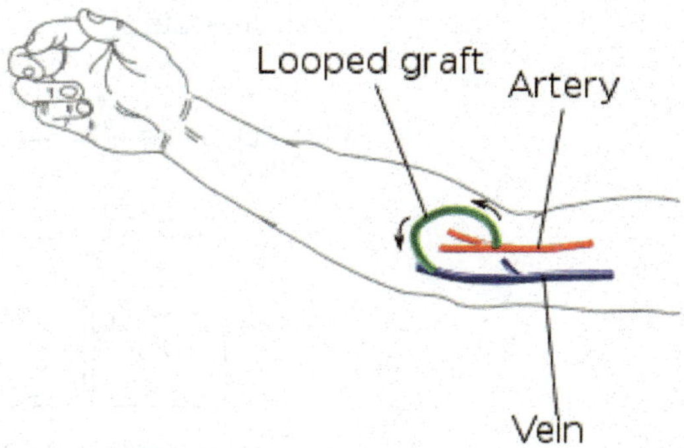

FIGURE 4.23: Synthetic Arteriovenous Vascular Graft

Kubiik / Wikimedia Commons / Copyright in the Public Domain.

This surgical procedure may be done a few weeks prior to the beginning of dialysis treatment. The small diameter vascular graft that is implanted can be fabricated from expanded polytetrafluoroethylene (ePTFE) or Dacron. Both materials are fairly hemocompatible and can be easily sutured to the artery and vein. As with the arteriovenous fistula, this method of vascular access provides additional sites for needle punctures and lessens the risks for blood clot formation. Once preparations of the patient's vasculature have been completed, patients follow their prescribed dialysis therapy schedules. Patients spend approximately three hours completing one hemodialysis cycle followed by an additional hour stabilizing their blood pressures. Significant volumes of fluid are removed from patients during sessions, thus reducing their blood pressures. Patients prescribed additional blood pressure–controlling medications must be particularly careful to maintain adequate blood pressures before, during, and after treatments. Hemodialysis therapy begins by accessing the patient's vasculature to allow arterial blood flow to the dialysis circuit and returning blood flow back into the patient's venous system (Figure 4.24) (57). Arterial blood is accessed from the patient by a needle and catheter system and is circulated through an arterial pressure monitor and blood pump that regulate the blood flow into the dialysis cartridge. All of the dialysis components, aside from the vascular access needle and initial portions of the catheter, are located on an

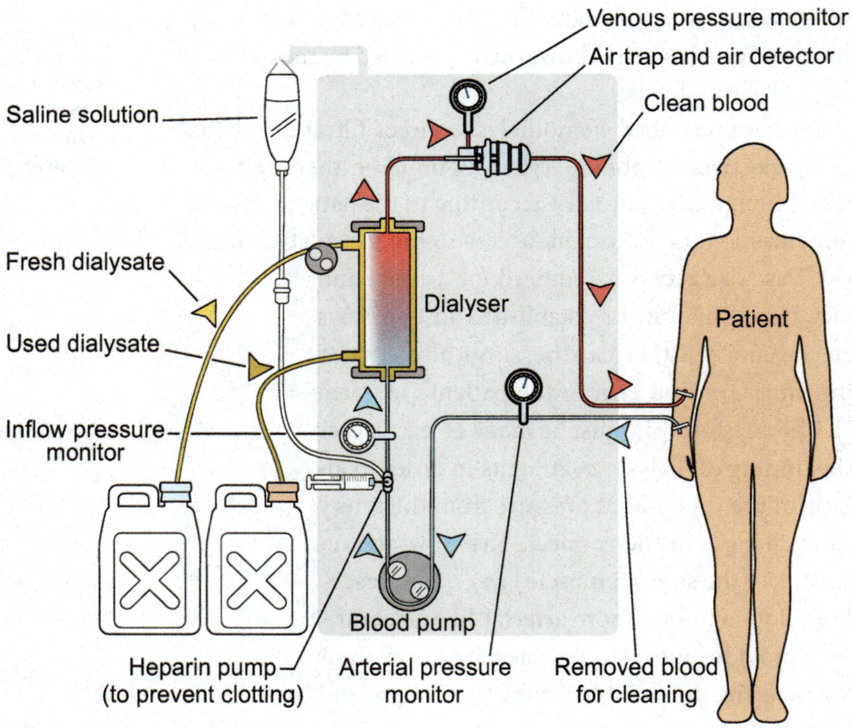

FIGURE 4.24: Dialysis Circuit

external console that is situated next to the patient's bed. Prior to entering the dialysis cartridge, blood is heparinized to prevent clot formation throughout the circuit. The dialysis cartridge contains approximately ten thousand hollow fiber membranes aligned in parallel (Figure 4.25). Each hollow fiber has an average diameter of 500 microns. The hollow fiber membranes possess a nominal molecular weight cutoff of 50,000 Daltons and are fabricated from either cellulose-based fibers or synthetic polymer fibers (58, 59). Molecules having a molecular weight of less than 50,000 Daltons are able to permeate through the pores of the membrane, whereas those with molecular weights greater than 50,000 Daltons remain behind in the blood. As blood passes from the catheter to the lumens of the hollow fiber membranes, small molecular weight entities such as excess salts, uremic toxins, and small proteins diffuse down their concentration gradients through the pores of the hollow fiber membranes and are removed by a countercurrent of dialysate circulating on the outside of the hollow fibers. Blood flow varies from 300–400 mL/min (Table 4.4) (60). Dialysate is fluid that has been formulated to carry the appropriate amount of solutes that will drive diffusion of excess solutes and toxins from the blood. Each patient is prescribed a specific dialysate formulation based upon their metabolic requirements. Dialysate flow rates are approximately 500–800 mL/min (60). As blood exits the dialysis cartridge through a venous catheter, it passes through a venous pressure monitor, an air detector, and an air trap and is then returned to the patient's venous circulation. In addition to removing excess solutes and uremic toxins, hemodialysis also removes a significant volume of fluid that has accumulated between sessions. This is accomplished by increasing the pressure of blood in the circuit prior to its flow through the dialyzer. Increased pressure results in ultrafiltration of water through the dialysis membrane and efficient fluid regulation. Each treatment session removes 1–3 liters of fluid and 50 ± 20 grams of uremic toxins (57, 60). Analysis of plasma urea nitrogen values over a one-week period demonstrates a sawtooth pattern, with high plasma urea nitrogen levels prior to the commencement of hemodialysis therapy and low plasma urea nitrogen values immediately following (Figure 4.26). These significant fluctuations in metabolites can deleteriously affect kidney and other organ functions.

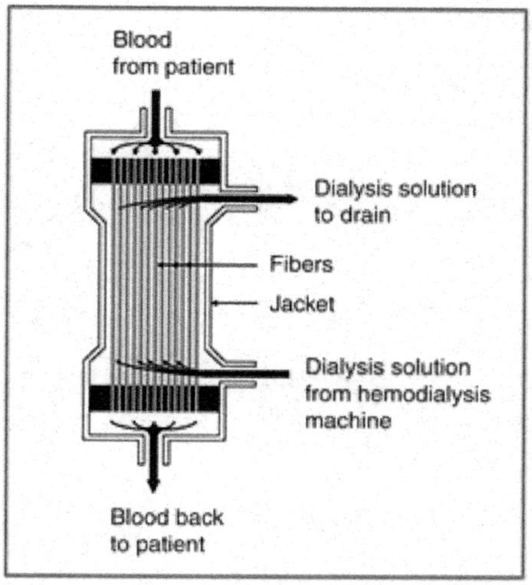

FIGURE 4.25: Hollow Fiber Dialyzer
NIDDK / NIH / Copyright in the Public Domain.

A second treatment option for patients suffering from chronic renal failure is peritoneal dialysis. Individuals who are not able to attend treatments at hemodialysis centers or who prefer to dialyze at home are prescribed daily peritoneal dialysis therapy. Removal of excess fluid and uremic toxins occurs through diffusional exchange between the peritoneal membrane and the abdominal cavity (Figure 4.27) (61). Prior to the initiation of treatment, patients are implanted with a percutaneous catheter that is capped between sessions. The catheter is the access port through which dialysate is delivered into and out of the abdominal cavity. At home, patients are trained to handle all components of the system in an aseptic manner. Treatments can be initiated in the

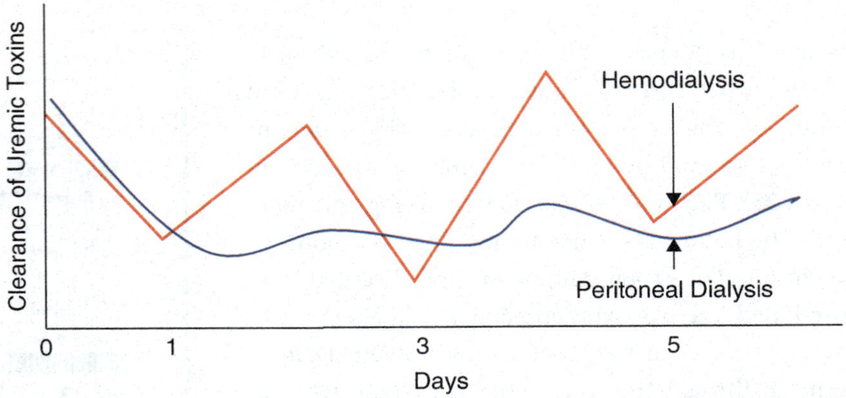

FIGURE 4.26: Theoretical Profile for Toxin Clearance: Hemodialysis vs. Peritoneal Dialysis (Original Figure)

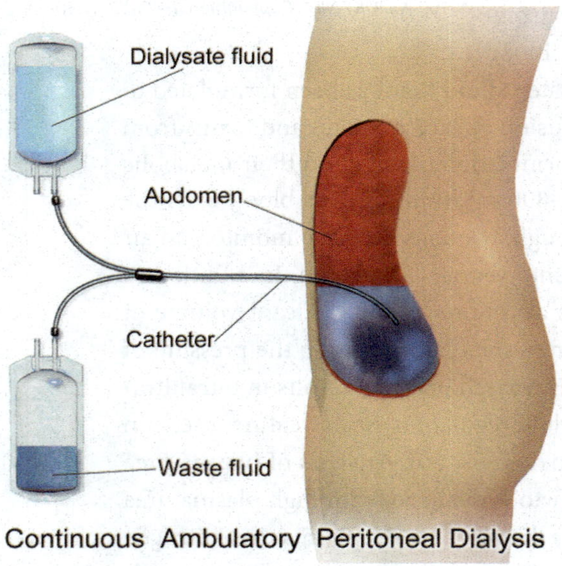

FIGURE 4.27: Continuous Ambulatory Peritoneal Dialysis

morning or evening and can either rely on gravity or a pump for delivering dialysate into and out of the peritoneal cavity. Commencement of treatment involves delivery of dialysate into the abdominal cavity through the percutaneous catheter. Once delivered, the catheter is capped and the dialysate remains in the abdominal cavity for up to eight hours. Diffusional exchange occurs over this time period (62). Excess solutes and uremic toxins diffuse out of the peritoneal capillaries and into the abdominal cavity, where they collect over the prescribed treatment period. The dialysate contains an excess of glucose that osmotically drives water from the capillaries and into the abdominal cavity as well. After six to eight hours, exchange is complete, and the dialysate is drained from the abdominal cavity. Treatment sessions occur at eight- to twelve-hour intervals. Although initially less expensive, peritoneal dialysis costs the patient as much as hemodialysis at the end of any given year due to expenses from retrofitting the patient's home and unexpected hospital stays resulting from infections. Infection, resulting from bacterial contamination of the percutaneous catheter, is the main adverse event associated with this process (63). Unlike hemodialysis, plasma urea nitrogen values remain fairly constant and within normal ranges due to daily maintenance with peritoneal dialysis (Figure 4.26).

c. Orthopedics and Prosthetics

i. Total Joint Replacement

The largest articular joint in the human body is the knee. The knee joint is a pivoting joint composed of the distal end of the femur, the femoral condyles, and the proximal end of the tibia, the tibial condyles, which are tethered together by the anterior, posterior, lateral, and medial ligaments (Figure 4.28). Several factors prevent direct bone-on-bone interactions of the femoral and tibial condyles during movement. The half-moon-shaped meniscus, composed of cartilage and connective tissue cells, is located on the tibial face and provides significant shock absorption and protection against high mechanical loads imparted on the joint space. Additionally, the femoral and tibial condyles are covered with thin layers of articular, hyaline cartilage tissues that reduce friction during motion. Articular cartilage is composed of specific cells called chondrocytes that produce predominantly type II collagen and chondroitin sulfate. The entire joint space is encapsulated in the synovial membrane. Specialized cells within the synovial membrane produce a viscous fluid—the synovial fluid—which functions to cushion the entire joint during motion and provides a diffusional substrate through which nutrients can circulate.

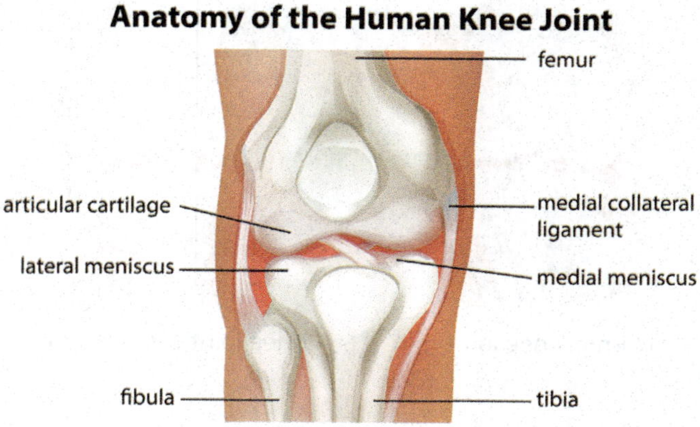

FIGURE 4.28: Anatomy of the Human Knee Joint
Copyright © 2013 by Depositphotos / blueringmedia.

The largest weight-bearing joint in the human body is the ball-and-socket hip joint. The replaceable components of the hip joint consist of the proximal end of the femur, the femoral head, and the pelvic socket, which is called the acetabulum (Figure 4.29). The femoral head rotates within the acetabulum, and both structures are secured in place by several large ligaments. Similar to the knee, a synovial membrane encapsulates the hip joint, and the synovial fluid reduces friction

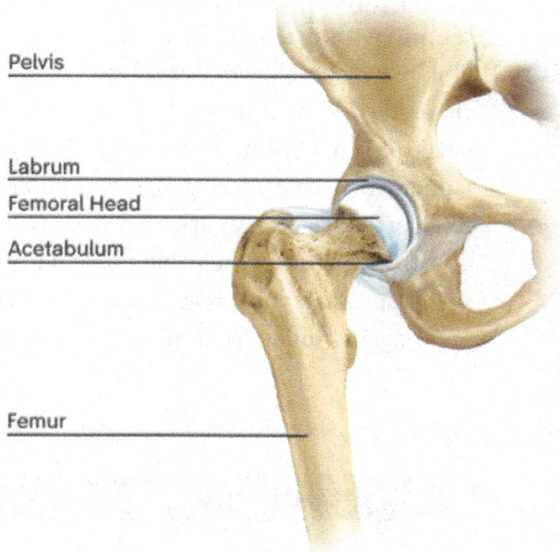

FIGURE 4.29: Anatomy of the Human Hip Joint
Copyright © 2011 by Smith & Nephew, (FAL 1.3) at: http://commons.wikimedia.org/wiki/File:Figure_1._Basic_anatomy_of_the_hip_joint.png. A full copy of the license can be found here: http://artlibre.org/licence/lal/en.

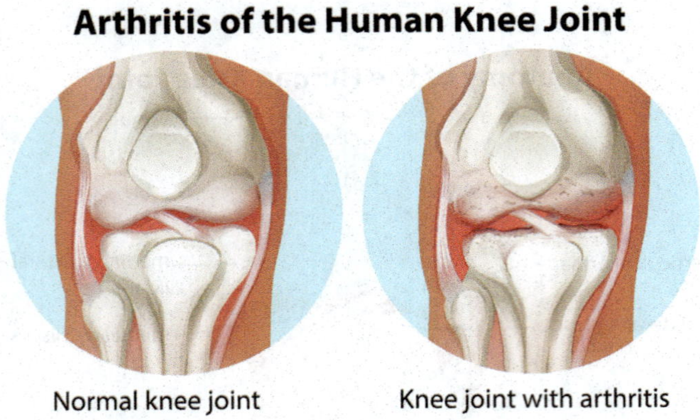

FIGURE 4.30: Arthritic Knee

during movement and provides a diffusional substrate for nutrients. Both the femoral head and the acetabulum are lined with thin layers of articular cartilage that prevents direct bone-on-bone interactions.

Inflammation and degeneration of the articular cartilage layers that line the bony ends of the hip and knee joints are progressive disorders cumulatively diagnosed as arthritis (Figure 4.30) (64). Eventual wearing of the cartilage results in painful bone-on-bone interactions and can potentially lead to necrosis of bone tissue. Arthritis afflicts approximately 32 million people in the United States and is second only to sinusitis as the most common chronic condition (65). Approximately 20 percent to 30 percent of people aged seventy and older suffer from osteoarthritis (OA) of the hip, and the total costs associated with arthritis are over $82 billion per year in the United States, including hospital and drug costs, nursing home costs, and lost productivity and work (65, 66).

Arthritis is a broad-spectrum disorder, and patients are typically diagnosed as either presenting with osteoarthritis, inflammatory arthritis, rheumatoid arthritis, or traumatic arthritis (Table 4.5). Osteoarthritis affects an estimated 21 million people in the United States and is characterized by mild to debilitating pain in the hands and joints such as knees, hips, feet, and back (65). According to the Arthritis Foundation, osteoarthritis is most prevalent among individuals forty-five years of age and older, and women tend to be more susceptible than men to this particular type of arthritis (67). Overall, osteoarthritis qualifies as the most common form of arthritis. Due to the progressive nature of cartilage degeneration and the inability of this tissue to spontaneously regenerate (particularly in older patients), the eventual pathology leads to bone-against-bone friction, pain, and eventual loss of movement. Symptoms include joint pain or

TABLE 4.5: Types of Human Arthritis (original table)

Type	Cause	Age of Presentation
Osteoarthritis	Breakdown of articular cartilage in joint space	>45 years
Inflammatory arthritis	Chronic inflammation of synovium resulting in degradation of articular cartilage in joint space	>20 years
Traumatic arthritis	Serious injury causes avascular necrosis of bone in joint space with resulting inflammation and degradation of articular cartilage	>1 year following initial injury

aching (often after exercise or extended periods of pressure on weight-bearing joints) and limited or eventual loss of range of motion.

Inflammatory arthritis is associated with a systemic immune response that elicits joint inflammation with resultant pain and swelling. Inflammation is the process by which the immune system attempts to resolve injury by prompting white blood cell movement into affected areas, resulting in engulfment of foreign debris and release of chemical agents that attract additional white blood cells. The process is cyclical and can either aid in resolving injury or continue on a pathological course causing chronic inflammation and further tissue damage. The etiology is often idiopathic, although rheumatoid arthritis, an autoimmune disease, is one of the principle pathologies linked to inflammatory arthritis. The synovial space that surrounds a joint is avascular; therefore, the delivery of nutrients to and removal of waste from articular cartilage occurs primarily by diffusion. The lack of vasculature in this protected space impedes repair and regeneration of cartilage tissue. As inflammatory cells and molecules slowly enter the synovium, fluid volumes within the synovial space increase, nerve endings become stimulated, articular cartilage degenerates, and pain results. In the case of rheumatoid arthritis, a cytokine, tumor necrosis factor, is overproduced and elicits a chronic inflammatory response in the joint.

The final general category of arthritis is traumatic arthritis caused by direct injury to the joint. Significant traumatic force imparted on joints can cause soft tissue damage, including cartilage, ligament, and meniscal damage, and subsequent induction of chronic inflammation. Trauma typically affects the underlying bone tissue as well, and damage to the vasculature of underlying bone results in bone cell death and tissue necrosis. Tissue necrosis induces acute inflammation that, if not resolved, progresses into chronic inflammation. As stated earlier, resolution of joint injury is dependent upon the appropriate inflammatory responses and tissue repair. If this fails to occur, then arthritis can result.

Several methods of first-order treatment are available for addressing all four categories of arthritis. The primary order of care involves nutritional and lifestyle changes such as weight loss in order to alleviate pain associated with inflammation. Secondary measures include pharmaceutical and biopharmaceutical measures that either reduce generalized inflammation, as with corticosteroids and aspirin regimens, or modify disease pathways, as with Enbrel®, which affects tumor necrosis factor binding in the joint space. Treatments are limited by efficacy and patient compliance. Surgical repair constitutes a final option for patients who have not responded sufficiently to disease management. Arthroscopy, partial joint replacement, and total joint replacement are three general options for patients considering surgical intervention.

Approximately 1.5 million arthroscopic procedures are conducted each year in the United States (68). This laparoscopic procedure costs $6,000 on average and can last from one to three hours (68). Patients range in age from twenty to sixty years old, although due to overuse injuries associated with adolescent and college sports, many more young adults have had to undergo arthroscopic procedures in order to remain competitive athletes. Arthroscopic procedures are typically used to remove pieces of loose bone and cartilage as well as to repair damaged tendons and ligaments, such as cruciate ligament tears, meniscal tears, and damaged joint surfaces

TABLE 4.6: Modular Components and Materials of Total Joint Replacements (original table)

Total Joint	Metal	Nonmetal	Polymer
Hip	Cobalt–chromium alloys, titanium alloys, stainless steel alloys, tantalum, carbon fiber composites	Ceramics: zirconium oxide, aluminum oxide	Ultra high molecular weight polyethylene
Knee	Cobalt–chromium alloys, titanium alloys, stainless steel alloys, tantalum, carbon fiber composites		Ultra high molecular weight polyethylene

(Figure 4.31) (69). Procedures are associated with very high success rates, although bleeding and infections may occur postoperatively.

Partial and total joint replacements are invasive procedures that replace affected tissues with modular components made of metal alloys and polymeric materials (Table 4.6) (70). Although patients primarily suffer from inflammation and degeneration of cartilage in their joint spaces, bone as well as cartilage is replaced during implantation. Patients receiving joint replacement implants are typically sixty-five years and older; however, younger populations are beginning to undergo joint replacement procedures in order to ultimately extend the time that they can be athletically active. Artificial knee replacement is the most common joint replacement procedure in the United States, with over six hundred thousand performed each year (71). Cost of the actual implant ranges from $4,000–$6,000 (71, 72). The cost of the entire procedure increases exponentially due to surgical expenses, in-hospital stay, and physical rehabilitation, which is critical for full recovery and functionality of the implant. Both knee and hip implants are modular and are sized for each patient based upon X-ray and magnetic resonance imaging (MRI) data (73). Knee implants consist of a femoral component and a tibial component (Figure 4.32). Stainless steel, cobalt-chrome, tantalum, and titanium alloys are

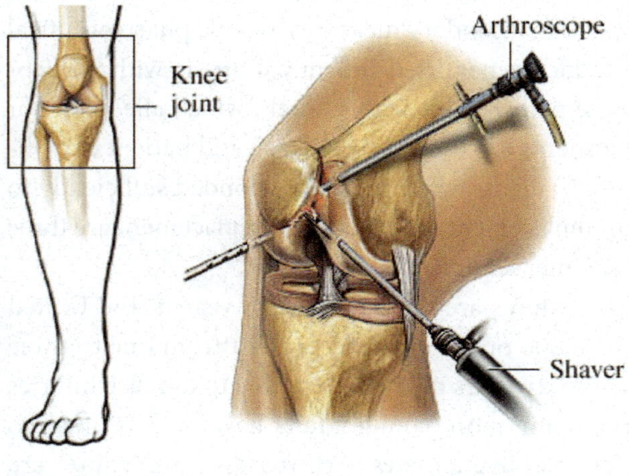

FIGURE 4.31: Arthroscopic Treatment of Arthritic Knee
Copyright © 2013 by Nucleus Medical Media, Inc. Reprinted with permission.

FIGURE 4.32: Knee Implant
Copyright © by Depositphotos / grieze.

metals that comprise the femoral component and the bone-anchoring stem of the tibial component. A high molecular weight polyethylene (HMWPE) spacer lines the portion of the tibial component that faces the joint space and articulates with the femoral component. This polymeric layer provides a smooth, frictionless surface that acts as a substitute for both cartilage and meniscal layers, thus providing reduction in heat generation as well as shock absorption upon mechanical impact. The knee is prepared for implantation by removing the femoral and tibial condyles and drilling into the proximal end of the tibia, thus removing the affected tissues. Fixation techniques for both the femoral and tibial components include press fitting by mechanical force, press fitting along with the addition of bone cement, and biological in-growth from apposing bone tissue. Lateral and medial ligaments may remain in place in order to stabilize the artificial knee, although this is dependent upon ligament health. Although total knee replacement has been the standard of care for the majority of patients, partial knee replacements that involve replacement of only half of the femoral condyle are available as well. Preservation of healthy cartilage by replacing only those areas exhibiting tissue damage preserves the integrity of the natural joint and allows for more effective revision surgeries in the future.

Artificial hip implantation is the second most common total joint repair surgery, with approximately three hundred thousand performed each year in the United States (74). Cost is similar to the total artificial knee, although there is more variability in available modular components (75). Surgical replacement of an affected hip can take anywhere from two to four hours and requires an in-hospital stay as well as postoperative physical rehabilitation. Patients typically receiving this type of implant either suffer from conditions such as rheumatoid arthritis or present with progressive osteoarthritis and osteoporosis due to advanced age. Similar to knee replacement, artificial hip components are sized for individual patients based upon MRI and X-ray data (73). During surgery, however, several different device sizes are available for implantation in case of unexpected differences in dimensions. The total artificial hip is composed of four modular pieces: the femoral stem, femoral head, acetabula cup, and acetabula liner (Figure 4.33). The femoral stem, femoral head, and acetabula cup can be fabricated from stainless steel, cobalt-chrome, tantalum, and titanium alloys. The acetabula liner is composed of HMWPE and provides a smooth surface upon which the femoral head can rotate within the acetabulum. Recently, total artificial hips have been implanted without the acetabula liner, thus allowing metal-on-metal interactions. Although initially considered a better option than metal or plastic implants, which have been associated with inflammation from accumulation of plastic wear debris within the joint space, metal-on-metal implants have also been fraught with adverse wear events (76). In 2013, Johnson & Johnson, Inc. was fined $4 billion on behalf of patients who suffered from the effects of the accumulation of metal wear debris within the joint space (77). The hip

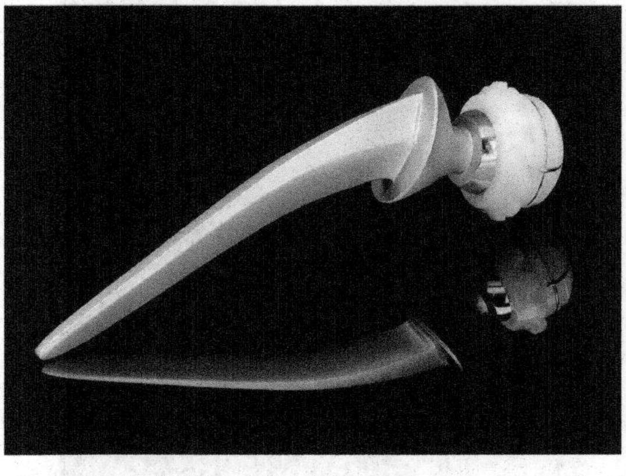

FIGURE 4.33: Hip Implant
Copyright © 2013 by Science Museum London / Science and Society Picture Library, (CC BY-SA 2.0) at http://commons.wikimedia.org/wiki/File:Stainless_steel_and_ultra_high_molecular_weight_polythene_hip_replacement_(9672239334).jpg.

FIGURE 4.34: Hip Implant with Porous Titanium Surface (arrow)
Copyright © by Depositphotos / leonidp.

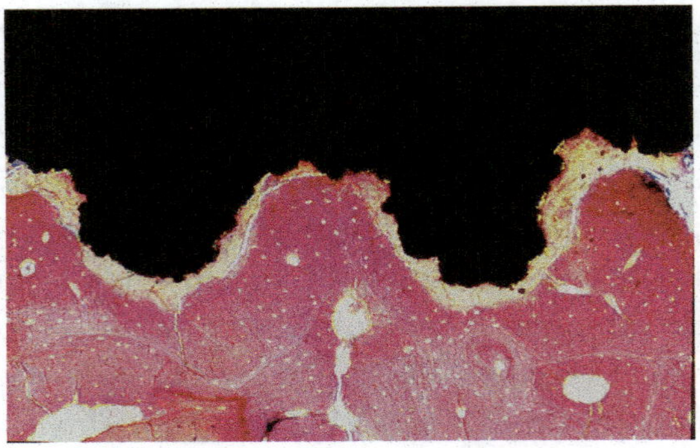

FIGURE 4.35: Bone Cell Integration of Titanium Alloy Surface
Robert Gougaloff / Copyright in the Public Domain.

is prepared for implantation by excising the femoral head and reaming the femoral shaft and acetabulum, thus removing the affected tissues. The femoral stem and acetabula cup can be fixed to the femoral and pelvic bones, respectively, by press fitting alone, press fitting with the addition of polymethylmethacrylate (PMMA) cement, and by bony integration with the surface of the implant. Osteointegration, bone cell integration, is the movement and growth of individual bone cells into the pores and trabeculae located on the surface of etched titanium alloy–based implants. Titanium alloy is biologically compatible material that has been shown to induce both movement, osteoconduction, and growth, osteoinduction, of bone cells. The surfaces of titanium alloy–based implants are etched to produce a trabecular surface containing pores ranging from 100–300 microns in diameter (Figure 4.34) (78). Once implanted, bone cells begin to migrate toward the implant surface and settle within the pores, where they produce collagen and become biologically fixed to the implant (Figure 4.35). Bone cell integration and implant fixation require approximately twelve weeks of non-weight bearing by the patient; therefore, patients follow a different postoperative therapy protocol than traditional implant patients. The goal of this particular type of fixation strategy is to seek a biologically active attachment for the lifetime of the implant.

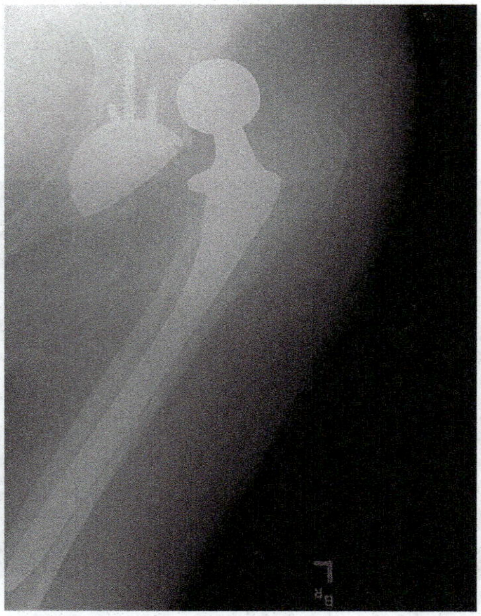

FIGURE 4.36: Dislocated Hip Replacement
Copyright © 2008 by Bill Rhodes, (CC BY 2.0) at https://commons.wikimedia.org/wiki/File:Dislocated_hip_replacement.jpg..

Although well established and successful, both knee and hip replacement surgeries can be associated with adverse postoperative events (79). Early postoperative events—such as dislocation and early infection—affect less than 10 percent of all patients (Figure 4.36). Dislocations are often the result of discrepancies between anatomical and device component sizes. Revision surgeries are typically necessary in order to replace the mismatched implant. Infections due to surgical contaminations are frequently treated by implanting a drain into the joint space and draining exudate from the infected area. Systemic antibiotics are also administered to patients. Device-associated infections are difficult to treat because bacteria deposit biofilm layers on device surfaces and are protected from antibiotic therapies once they have colonized within the biofilm. If infections are detected prior to this event, then implants can be salvaged. If infections cannot be resolved, revision surgeries are necessary to replace the affected devices. Late-stage events occurring five years postoperatively and beyond include wear of articular-bearing surfaces; osteolysis, or bone cell death; mechanical loosening of the implant; and periprosthetic fractures along bone that is juxtaposed to the device. Both metal-on-metal implants and metal-on-plastic implants have shown evidence of wear debris accumulation in the joint space (80). Microscopic fragments of either plastic or metal induce chronic inflammation within the joint, which leads to destruction of surrounding tissues, stimulation of nerve endings, and pain. Incompatibility of device materials with surrounding bone tissues also leads to induction of inflammatory responses and potentially loosening of the implant on a microscopic scale (Figure 4.37). Microscopic loosening, called micromotion, results in further propagation of inflammatory responses and accumulation of fluid around the implant (80). Finally, periprosthetic fractures along bone tissues that are juxtaposed to the implant typically result from stress shielding of these tissues by device components. Mechanical loads—necessary for healthy bone growth—can be shielded from periprosthetic tissues and redirected to implants that are composed of materials that make them mechanically too stiff and unyielding. In this environment, healing bone tissue is prevented from mechanical stress exposure and cannot build adequate bone density. Decreased bone density results in secondary fracturing. Late-stage events, as mentioned above, most often require revision surgeries, repair of affected tissues, and replacement with new devices.

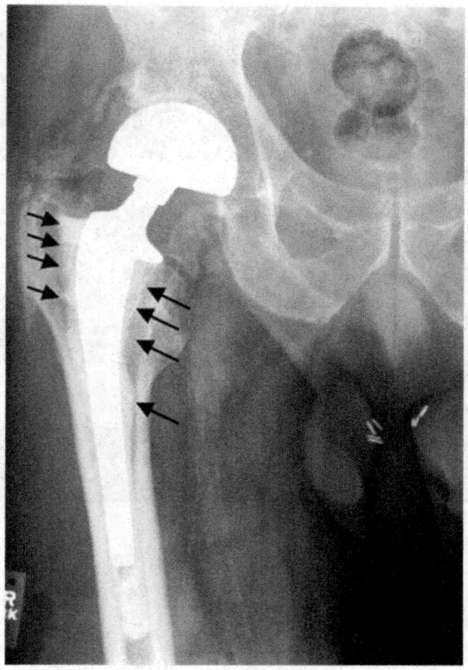

FIGURE 4.37: Aspeptic Loosening of Hip Implant

ii. Upper and Lower Limb Prosthetics

As of 2014, approximately 1.7 million people in the United States have been afflicted with upper (10 percent) or lower (90 percent) limb amputations (81). The science of replacing an amputated limb with either a nonfunctional or functional device is known as prosthetics. Upper- and lower-limb prosthetic devices are in high demand in both developing and developed countries,

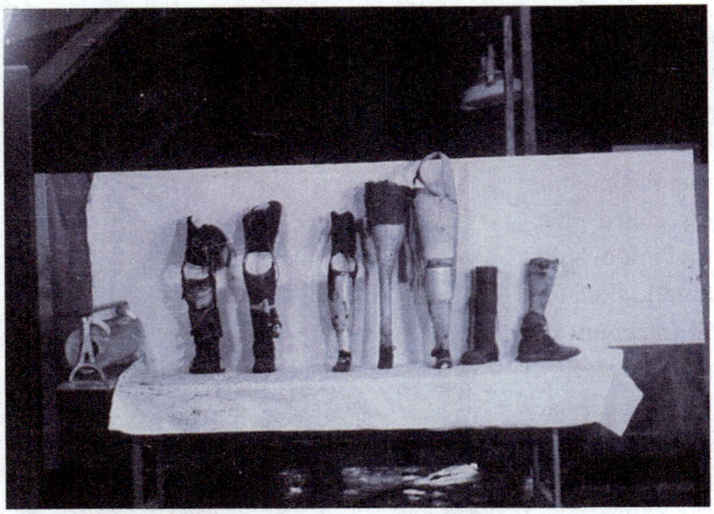

FIGURE 4.38: Early Lower Limb Prostheses
NIH / Copyright in the Public Domain.

albeit for significantly different reasons. Amputations in developing countries are primarily the result of traumatic injuries from industrial and traffic accidents, military conflicts, and resultant land mine explosions. Circulatory insufficiencies resulting from diabetes and atherosclerotic blockages, in addition to congenital defects, are responsible for the majority of amputations performed in developed countries. However, over the past ten years in the United States, a significant rise in the number of amputations has been attributed to traumatic injuries suffered by troops in the Afghanistan and Iraq wars.

The use of prosthetic devices dates back to ancient Egypt and, although useful for filling anatomical spaces left behind by the amputations, these devices, for the most part, were nonfunctional (Figure 4.38). In fact, until the latter part of the twentieth century, both upper- and lower-limb prosthetics were ineffectual, causing significant stress on the amputee. In 1945, the National Academy of Sciences established the Artificial Limb Program in response to the influx of World War II veteran amputees (82). Distribution of governmental funds to support this program resulted in the advancement of prosthetic design and function, including modifications for patient comfort, weight, ease of attachment, durability, functionality, and cosmetic appearance. Many current research programs in industry and academia are also being funded by governmental agencies such as the Defense Advanced Research Project Agency.

Upper- and lower-limb devices are categorized as transtibial, below the knee; transfemoral, above the knee; transradial, below the elbow; and transhumeral, above the elbow. Inclusion of manual or computerized joints in transhumeral and transfemoral prosthetics adds a level of complexity to device design and enhances the levels of utility. Costs for multifaceted prosthetics are significantly higher than for traditionally simple designs. The basic components of prosthetic limbs include the pylon, the socket, and the suspension system (Table 4.7) (83). The pylon consists of the internal frame of the prosthetic. Traditional pylons have been fabricated from metal-alloy rods such as stainless steel, which are heavy and cumbersome. Modern frames can be fabricated from lightweight carbon fibers that are strong and flexible and can be covered by materials that mimic the aesthetics of human skin (Figure 4.39, Figure 4.40) (84). The second component, the prosthetic socket, is the portion of the device that interfaces with the patient's residual stump and is typically composed of thermoplastic polymers such as polypropylene or polyethylene. These polymers are molded at high temperatures and return to their solid-state conformations once cooled. As a result, sockets fabricated from these materials can be designed to fit snugly against

TABLE 4.7: Basic Components of a Prosthetic Device (original table)

Component	Description
Pylon	The pylon consists of the internal frame of the prosthetic. Traditional pylons have been fabricated from metal-alloy rods, such as stainless steel, which are heavy and cumbersome. Modern frames can be fabricated from lightweight carbon fibers that are strong and flexible and can be covered by materials that mimic the aesthetics of human skin.
Socket	The socket is the portion of the device that interfaces with the patient's residual stump and is typically composed of thermoplastic polymers such as polypropylene or polyethylene. These polymers are molded at high temperatures and return to their solid state conformations once cooled.
Suspension	The suspension system is the mechanism by which the prosthetic is attached to the body and is designed in the form of a harness, strap, or suction.

the residual stump. Although soft liners like silicone are used as buffering materials between the socket and skin covering the residual stump, many patients experience skin sensitivity and irritation, which can lead to discomfort and damage to skin and underlying tissues (85, 86). The last basic component of a prosthetic is the suspension system. The suspension system is the mechanism by which the prosthetic is attached to the body and is designed in the form of a harness, strap, or suction. Although all prosthetics, upper and lower limb, consist of these three basic components, each patient is custom fitted for his or her prosthetic, and the devices are built on an individualized basis. Basic upper- and lower-limb prosthetics with limited functions can cost between $50,000 and $80,000, whereas more advanced devices that include significant ranges of motion and dexterity may cost over $100,000 (87).

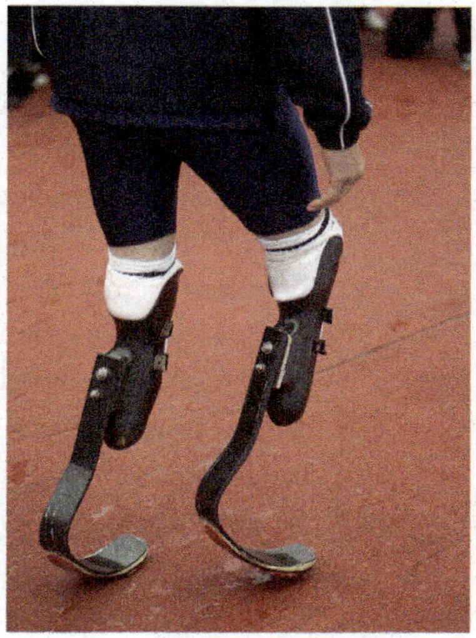

FIGURE 4.39: Carbon Fiber Prosthetic Lower Limbs

Over the past fifty years, upper-limb devices, although constituting less than 10 percent of limb prosthetics, have significantly increased in complexity and performance due to the intricacies required of natural hand and digit functions. These upper-limb devices can be divided into body powered, or cable controlled, and myoelectrically controlled devices (Figure 4.41) (88, 89). Cable-controlled prostheses enable the amputee to move residual muscle groups located proximal to the amputated stump in order to control movement and force of the terminal device. Exterior cables lead from residual muscle mass to distal motorized components of the prosthetic hand and digits. Changes in cable lengths result in initiation of movements in the distal prosthetic limb with resultant opening and closing of the clawlike prosthetic hand. This category of device has limited range of motion and prehensile strength. Modern myoelectric

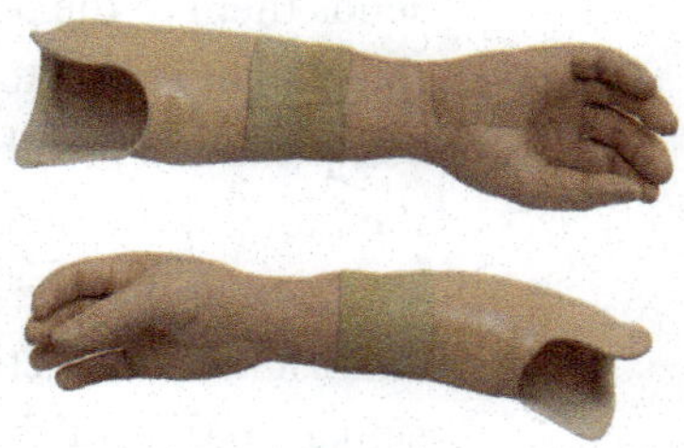

FIGURE 4.40: Skin-Like Covering for Prosthetic Arm

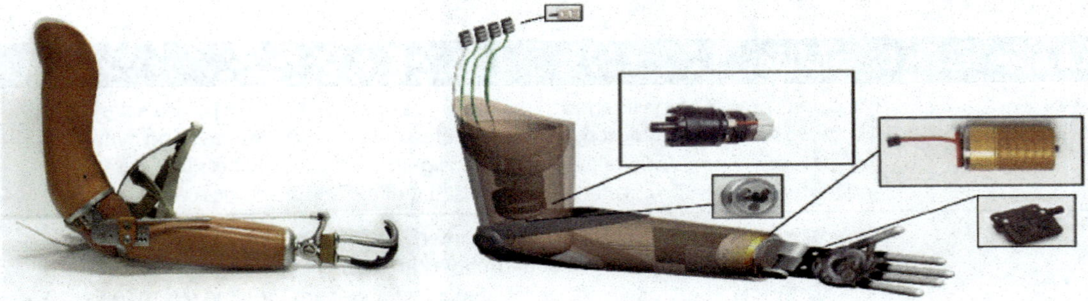

FIGURE 4.41: a) Cable-Controlled Upper Limb (Prosthetic) b) Myoelectric Controlled Prosthetic

devices employ electrical signals elicited by contracted muscles located near the amputated stump in order to activate movement of the distal prosthetic hand and digits. Surface-mounted electrodes detect electrical signals of selectively contracted muscle groups, which are then transduced and carried down to motors located in the prosthetic hand and digits. Depending upon the number of motors, amputees can open and close either a simple clawlike grip or manipulate each digit independently while manually manipulating the prosthetic thumb. Upper-limb prosthetics have been further modified to provide basic sensory information when grasping objects. This allows the patient more accurate motor control when performing tasks. Advanced lower-limb designs include microprocessors and sensors that measure angles and forces of a patient's gait. Computer-integrated knees and ankles provide additional feedback that allows patients to navigate small changes in terrain fairly easily. One of the most innovative prosthetic designs—targeted muscle reinnervation—involves redirection of the nerves from the amputated limb to an ectopic bundle of healthy muscle (Figure 4.42) (90, 91). Nerves from the amputated limb are anastomosed to several muscle bundles in the patient's chest. Patients are trained to selectively contract chest muscles that have electrodes from the prosthetic attached. As patients selectively contract their chest muscles, electrical signals generated in the nerves are transferred to a microprocessor unit that transduces the signals and sends them to the prosthetic. By thinking of a movement and contracting the appropriate muscles, the patient is able to move the prosthetic with a substantial amount of dexterity. The next level of complexity in prosthetic development, neural interfacing, involves direct communication between the brain and device by connecting nerves in the patient's motor cortex to the prosthetic (Figure 4.43). Neural interfacing involves implantation of microelectrodes in portions of the brain that control motor function. When patients think of a motor command, microelectrodes intercept the signals and send them directly to the prosthetic device. Currently, large-animal trials are being performed to test hard-wired circuits. Wireless circuits are the next step.

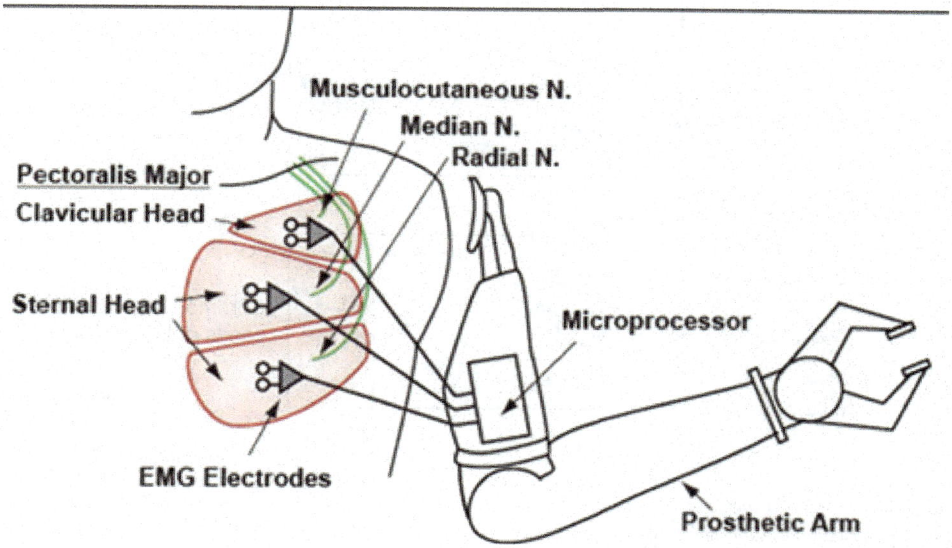

FIGURE 4.42: Targeted Muscle Reinnervation
U.S. Department of Veterans Affairs / Rehabilitation Research & Development Service / Copyright in the Public Domain.

d. Soft Tissue Repair and Replacement

i. Meniscal Reconstruction

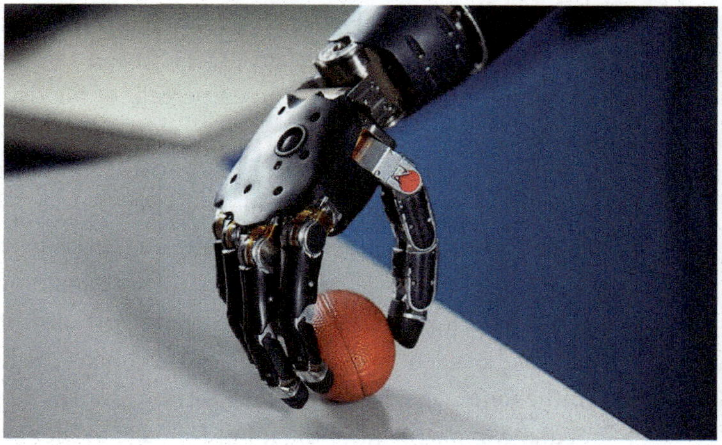

FIGURE 4.43: Brain-Controlled Prosthetic Hand
FDA / Copyright in the Public Domain.

Meniscal tissues are critical for healthy knee function. These tissues afford the knee joint with load sharing, shock absorption, reduction of joint contact stresses, increase in joint congruity and contact area, articular cartilage nutrition, passive stabilization, and limitations of extreme flexion and extension. Although the menisci are the physiological targets of dehydration and degeneration in older patients, younger patients are now presenting with damaged menisci due to crushing and tearing during traumatic injuries associated with competitive athletics. Surgery and repair options vary depending upon the location and extent of tissue damage, and removal of 50 percent or greater of meniscal tissue results in long-term joint damage with resultant arthritis.

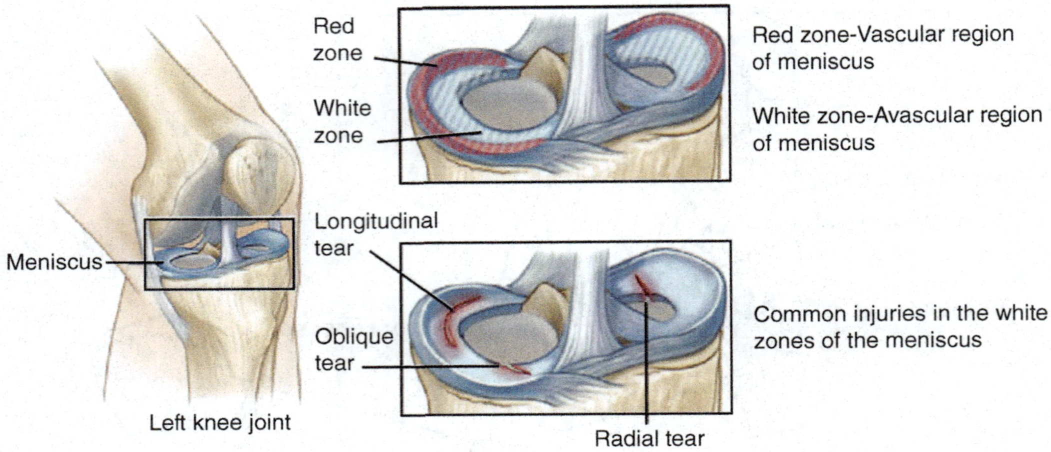

FIGURE 4.44: Meniscus and Meniscal Zones

The menisci are soft fibrous connective tissues that are located on the tibial plateau of both knees and are composed of 72 percent water, 60 percent to 70 percent of types I and II collagen (dry weight), and 2 percent to 3 percent glycosaminoglycans (GAG; dry weight) (Figure 4.44). The cellular components are graded with fibroblasts primarily located at the peripheral edges (red-red zone), fibrochondrocytes located adjacently (red-white zone), and chondrocytes located medially (white-white zone) (Figure 4.44). Circumferential and radial collagen fibers, synthesized by the fibroblasts, make this tissue anisotropic and ensure strength in three dimensions. Blood is supplied to the peripheral 20 percent to 25 percent of the tissue via the perimeniscal geniculate plexus. As a result, damaged tissues located on your near-to-the-peripheral zones of the meniscus have the potential to repair due to adequate vascularization that allows immune cell recruitment and promotion of reparative processes. Injuries to the medial portions, however, are not able to be resolved because of that zone's avascular nature and inability to recruit reparative cells to that location.

The meniscus is divided into two distinct anatomical structures, with the medial portion of the meniscus being *C*-shaped and the lateral portion being circular. Both components are attached to the tibial plateau by ligamentous connections. The lateral portion absorbs 70 percent of the load on the joint, whereas the medial portion absorbs 50 percent of the force transmitted through the medial compartment of the joint. Damage to meniscal tissues occurs in two planes: longitudinal and horizontal (92). Longitudinal injuries are categorized as complete and incomplete buckle handles, oblique, and transverse. Horizontal injuries are usually degenerative in nature. The most frequent presentations of meniscal injuries are mechanical symptoms, including instability and pain with deep knee bending. Along with physical examination, MRI is the diagnostic tool most often used to make a definitive diagnosis.

Meniscal repairs were first described in the late nineteenth century by Thomas Annandale, but it was not until the early 1980s that clinically relevant arthroscopic repairs were introduced (93). Surgical options include partial and full meniscectomies; arthroscopic suturing with nonresorbable

TABLE 4.8: Meniscal Repair Options (original table)

Injury Zone	Clinically Used	Clinically Used	Experimental	Experimental
Red	Self-healing	Sutures, mechanical Fixation devices, allograft transplantation	Xenogeneic collagen and intestinal submucosa implantation	Tissue-engineered implant, PLA/PGA, stem cell implantation
White	Partial or full meniscectomy			Tissue-engineered implant, PLA/PGA, stem cell implantation

sutures; arthroscopic fixation with anchors, tacks, arrows, and staples; allogeneic tissue transplantation; and novel approaches that use reconstructed xenogeneic meniscal transplants (Table 4.8) (94). Partial meniscectomies are performed arthroscopically. Portions of frayed tissues are trimmed, taking care not to remove the attached meniscal rim. Questions remain as to whether the existing tissue has the potential to partially regenerate. In most clinical cases, tissues remaining after partial meniscectomies progressively degenerate and eventually require full excision. Full meniscectomies can be performed arthroscopically or as open procedures. Once a majority of the meniscus—specifically the lateral meniscus—has been removed, studies have shown that contact forces on the articular cartilage increase by approximately 300 percent (95). Chronic focal contact forces on the articular surfaces of the femoral condyles lead to articular cartilage degeneration and arthritis. Progressive articular cartilage degeneration ultimately results in the need for total joint replacement, as discussed in the previous section. Injuries that qualify for surgical repair are most often resolved using the gold standard technique for repairing tissue in the vascular and semivascular zones: arthroscopic fixation with nonresorbable sutures (94). The anterior cruciate ligament (ACL) is often replaced during this procedure, and studies have shown lower healing rates without concomitant ACL repair. Mechanical fixation devices such as arrows, staples, and tacks can also be used to repair partial tears. Devices can be fabricated from resorbable and nonresorbable materials such as polylactic acid/polyglycolic acid (PLA/PGA) copolymer blends and braided polyesters, respectively. Adverse events are associated with all types of mechanical meniscus repair. Although nonresorbable sutures have the greatest mechanical strength, arthroscopic knot tying is difficult, and proper tissue fixation can be compromised. An additional concern with suture repair of posterior regions of the meniscus is neurovascular injury. Materials-based fixation devices have lower incidences of neurovascular injury, but they have been correlated with chronic inflammation and foreign body giant cell (FBGC) responses. FBGC responses are fairly common with all implantable synthetic devices and involve recruitment of phagocytic immune cells such as macrophages to the device surface. Once at the surface, macrophages persist and eventually fuse to form giant cells that encase the implant. Material-based fixation devices also have lower mechanical strengths than sutures, cost more, and have been shown to migrate through the remaining meniscal tissue.

A final clinical option for patients with meniscal damage is transplantation of allograft meniscal tissue (96). Indications for allograft transplantation include those patients who have had full meniscectomies, patients undergoing concomitant ACL reconstruction and those with progressive osteoarthritis. Allograft tissues, those tissues transplanted between individuals of the same species, can be fresh, cryopreserved, or freeze-dried prior to transplant. Transplantation is an open procedure and positioning of the grafted tissue is critical for successful functioning. Grafts can be sized directly on the surgical field but manipulation and fixation have proven to be difficult. Other issues associated with transplantation are availability of viable tissue, sizing, biological variability between samples, potential for disease transmission, survivorship, and variations in surgical techniques and placement. Since these transplants are acellular and do not carry immune-stimulating proteins, immunosuppressants are not necessary, although inflammatory responses can occur due to the surgical procedure. Novel xenogeneic meniscal constructs are currently being developed and tested in preclinical models (97). Xenogeneic tissues such as processed bovine Achilles tendon and porcine small intestinal submucosa are harvested, minced, and processed in enzymes, salt, and acidic washes in order to remove cellular components. Once purified, the tissues are lyophilized (freeze-dried) and then molded into appropriate shapes and sizes. Implants are only feasible for replacing portions of the meniscus in red-red or red-white zones. The entire meniscal rim must be intact and unstable segments must not be present. Concerns with xenogeneic transplantation include transmission of zoonotic diseases, long-term stability of the transplant and surrounding tissues, and proper integration of host tissues. Currently, meniscal tissue has proven to be one of the most difficult and challenging soft tissues to repair, and future studies focusing on synthetic replacement of the entire tissue is now underway.

Technology on the Edge: The Bioartificial Kidney

The incidence rate of end-stage renal disease (ESRD) in the United States has increased over the past twenty-five years. ESRD is commonly associated with comorbidities such as diabetes type 2 and heart disease, and the incidence rates of both diabetes type 2 and heart disease are increasing as well. Most patients presenting with chronic kidney failure are prescribed either hemodialysis or peritoneal dialysis. Although peritoneal dialysis maintains better physiological levels of glucose and insulin, complications related to infection are often the reasons that physicians refer to hemodialysis for their patients. For decades, scientists and physicians have attempted to develop renal support systems that provide additional benefits to patients without the common side effects associated with traditional hemodialysis. Patients prescribed weekly regimens of hemodialysis experience significant plasma fluctuations in uremic toxins, fluid retention, and fluctuations in blood pressure. Hemodialysis sessions often take up to five hours to complete and are extremely disruptive to patients' daily lives.

An alternative to weekly hemodialysis is development of the bioartificial kidney (98). Unlike traditional dialysis, which simply removes uremic toxins and maintains fluid balances, the bioartificial kidney is intended to provide continuous metabolic support and enhanced biochemical profiles (99, 100). Initial attempts at creating a bioartificial kidney focused on paracorporeal devices comprised of conventional polymeric hollow fibers and xenogeneic renal tubule cells (101). These early prototypes demonstrated proof of principle and laid the groundwork for fully implantable systems. Implantable bioartificial kidney systems combine the fluidic profiles of peritoneal dialysis with hollow fiber designs of traditional hemodialysis in the attempts to maintain normal physiological profiles of uremic metabolites and provide a large surface area for maximized molecular exchange (102). Xenogeneic cells are seeded along the inner surface of polysulfone hollow fibers, and the entire system is polymerically encased to provide protection against rejection by recipients. Seeded cells eventually form confluent layers that metabolically function as normal renal tubule cells, and once implanted, the device can be anastomosed to arterial and venous conduits that allow blood flow through the systems (103). Ultimate success for the implantable bioartificial kidney relies on the diffusive and convective abilities of the bioartificial construct as well as viability of the encased xenogeneic cells.

End-of-Chapter Questions and Answers

1. What is the primary underlying pathology that results in cardiovascular disease in the United States?

   Diabetes
2. Which type of coronary arterial plaque is *primarily* composed of white blood cells, cholesterol, and platelets?

   Atheroma
3. What are the primary percutaneous interventions used to treat coronary arterial blockages?

   Balloon angioplasty and stenting
4. What is the physiological process that involves growth of smooth muscle cells and secretion of extracellular matrix into the lumen of a percutaneously treated coronary artery?

   Neointimal hyperplasia
5. What is the clinically relevant term that is used to describe the secretion of extracellular matrix into the lumen of a percutaneously treated coronary artery along with reactive elastic recoil of the affected vessel?

   Restenosis
6. Which specific chemotherapeutic drug was incorporated into the first drug-eluting coronary stent to prevent subsequent closure of the affected vessel?

   Taxol
7. What is the maximum number of arteries that can be bypassed in a MIDCAB surgery?

   One to two
8. What is the maximum number of arteries that can be bypassed in a CABG surgery?

   Four to five

9. Which of the three coronary artery bypass procedures requires extracorporeal membrane oxygenation (ECMO)?
   CABG
10. What is the term that describes cardiac valve failure resulting from the accumulation of calcium on valve leaflets and stiffening of the valve?
   Valvular stenosis
11. From which tissues are the cardiac valves composed?
   Collagen and fibroblasts/connective tissues
12. Which type of replacement valve would a seventy-three-year-old patient, who is already maintained on anticoagulants, most likely receive?
   A mechanical valve
13. What are the two types of *device* failures associated with tissue-based cardiac valves?
   Fatigue and tearing
14. What is the primary source of *device* failure associated with mechanical cardiac valves?
   Thrombosis
15. Which cardiac chamber is most often supported by a ventricular assist device?
   Left
16. What are the three clinical uses of a left ventricular assist device?
   a. **Bridge to recovery**
   b. **Bridge to transplant**
   c. **Destination therapy**
17. What is the most frequent complication that patients experience following implantation of their VADs?
   Neurological complications
18. How many liters of fluid do the kidneys filter in a twenty-four-hour period?
   180 liters
19. Why are more kidneys transplanted than any other organ each year?
   Kidneys can be donated by living donors
20. What diagnostic measurement is most commonly used by physicians to determine kidney overall function?
   Clearance rate
21. Glomerular filtration rate (GFR) is often used to determine kidney filtration function. What substance is used to test GFR?
   Inulin
22. Vascular access for dialysis therapy occurs by two means. What are the two types of access?
   a. **Arteriovenous fistula**
   b. **Synthetic vascular graft**
23. How is excess fluid removed during dialysis sessions?
   Ultrafiltration

24. In what way do the daily plasma urea levels for a patient receiving peritoneal dialysis differ from those of a patient receiving hemodialysis?

    Urea levels increase between dialysis treatments but remain within the normal range.

25. Which adverse event is most often associated with peritoneal dialysis?

    Infection

26. Why do the synthetic membranes used in hemodialysis become less efficient over time?

    Fouling or clogging of the pores following initial exposure to blood components

27. Which minimally invasive procedure is typically used to remove debris from the synovial joint space?

    Arthroscopy

28. Which type of arthritis is specifically associated with autoimmunity?

    Rheumatoid arthritis

29. What high-density plastic is used as the articulation surface in artificial knees and most artificial hips?

    Ultrahigh molecular weight polyethylene

30. What is the most prevalent early-stage adverse event associated with both total knee and total hip replacement?

    Infection

31. Which late-stage adverse event is associated with unequal load sharing between the implant and native bone?

    Micromotion

32. Which metal alloy can be etched to produce a porous surface and is used in the construction of both artificial knee and hip implants?

    Titanium alloy

33. In developing countries, which disease is most often associated with lower-limb amputation?

    Diabetes

34. Which portion of a prosthetic is associated with soft tissue irritation, damage, and discomfort?

    Socket

35. Which type of prosthetic functions via contraction of muscles in the residual stump, electrical signal transduction, and electrical motor control?

    Myoelectric

36. What cutting-edge technology involves the direct connection between the cerebral motor cortex and prosthetic device resulting in mind controlled prostheses?

    Neural interfacing

37. Which portion of the meniscus experiences the most successful rates of natural healing following damage?

    The portions of both the medial and lateral menisci that are associated with the perimeniscal capillary plexus

38. What is the most common long-term complication associated with complete meniscectomy?
    Development of arthritis
39. In what type of meniscal injury are sutures and mechanical fixation devices most often used?
    Injuries in the vascular region
40. Meniscal allograft tissue may be processed in three ways prior to transplantation. What are the three ways?
    Fresh, freeze-dried, and cryopreserved tissue
41. For menisci that qualify for surgical repair, which procedure is considered the "gold standard?"
    Repair with nonresorbable sutures
42. Which host reaction is most often associated with materials-based fixation devices?
    Foreign body giant cell reactions

Works Cited

1. Go, A. S., D. Mozaffarian, V. L. Roger, E. J. Benjamin, J. D. Berry, M. J. Blaha, S. Dai, et al. "Heart Disease and Stroke Statistics—2014 Update: A Report from the American Heart Association." *Circulation* 129.3 (2014): E28–92.
2. Lloyd-Jones, D., R. J. Adams, T. M. Brown, M. Carnethon, S. Dai, G. De Simone, T. B. Ferguson, et al. "Heart Disease and Stroke Statistics—2010 Update: A Report from the American Heart Association." *Circulation* 121.7 (2010): E46–215.
3. Micha, R., S. K. Wallace, and D. Mozaffarian. "Red and Processed Meat Consumption and Risk of Incident Coronary Heart Disease, Stroke, and Diabetes Mellitus: A Systematic Review and Meta-Analysis." *Circulation*121.21 (2010): 2271–283.
4. "History of the Heart." *The Human Heart: An Online Exploration from The Franklin Institute, Made Possible by Unisys*. Franklin Heart Institute, 2014. http://learn.fi.edu/learn/heart/history/history.html.
5. Oppenheimer, Gerald M., MD, PhD. "Becoming the Framingham Study 1947–1950." *American Journal of Public Health* 95.4 (2005): 602–10.
6. Bitton, Asaf, MD, and Thomas Gaziano, MD MSc. "The Framingham Heart Study's Impact on Global Risk Assessment." *Progress in Cardiovascular Diseases* 53.1 (2010): 68–78.
7. "Heart Disease." *Centers for Disease Control and Prevention*. National Center for Chronic Disease Prevention and Health Promotion, Division for Heart Disease and Stroke Prevention, 05 Feb. 2014. www.cdc.gov/heartdisease/.
8. Diodato, Michael, and Edgar G. Chedrawy. "Coronary Artery Bypass Graft Surgery: The Past, Present, and Future of Myocardial Revascularisation." *Surgery Research and Practice* 2014 (2014): 1–6.
9. King, S. B. "Angioplasty from Bench to Bedside to Bench." *Circulation* 93.9 (1996): 1621–629.

10. Schuijf, J. D. "Cardiac Imaging in Coronary Artery Disease: Differing Modalities." *Heart* 91.8 (2005): 1110–117.
11. "Coronary Artery Bypass Graft Surgery." www.surgeryencyclopedia.com/Ce-Fi/Coronary-Artery-Bypass-Graft-Surgery.html.
12. Varieur Cadet, Justine. "AHA: PCI Is More Costly than CABG for Diabetics with Advanced CAD in the Long Run." *Cardiovascular Business* (2012).
13. Verma, S. "Off-Pump Coronary Artery Bypass Surgery: Fundamentals for the Clinical Card iologist." *Circulation* 109.10 (2004): 1206–211.
14. "Heart Bypass Surgery—Minimally Invasive." *New York Times*, 25 Jul. 2014, Health Guide sec.
15. Goldman, S., K. Zadina, and T. Moritz. "Long-term Patency of Saphenous Vein and Left Internal Mammary Artery Grafts after Coronary Artery Bypass Surgery." *ACC Current Journal Review* 14.4 (2005): 50.
16. Ryan, Thomas, MD. "The Coronary Angiogram and Its Seminal Contributions to Cardiovascular Medicine Over Five Decades." *Circulation* 106 (2002): 752–56.
17. Payne, Misty M. "Charles Theodore Dotter: The Father of Intervention." *Texas Heart Institute Journal* 28.1 (2001): 28–38.
18. Hurst, J. Willis. "The First Coronary Angioplasty as Described by Andreas Gruentzig." *American Journal of Cardiology* 57.1 (1986): 185–86.
19. Forssmann-Falck, Renate. "Werner Forssmann: A Pioneer of Cardiology." *American Journal of Cardiology* 79.5 (1997): 651–60.
20. Epstein, A. J., D. Polsky, F. Yang, L. Yang, and P. W. Groeneveld. "Coronary Revascularization Trends in the United States, 2001–2008." *JAMA: Journal of the American Medical Association* 305.17 (2011): 1769–776.
21. Farooq, Vasim, MBChB, MRCP, Bill D. Gogas, MD, and Patrick W. Serruys, MD, PhD. "Restenosis Delineating the Numerous Causes of Drug-Eluting Stent Restenosis." *Circulation: Cardiovascular Interventions* 4 (2011): 195–http://circinterventions.ahajournals.org/content/4/2/195.full.
22. Dangas, G. "Restenosis: Repeat Narrowing of a Coronary Artery: Prevention and Treatment." *Circulation* 105.22 (2002): 2586–587.
23. Massachusetts General Hospital. Cardiology. *Choice to Use Drug-eluting Stents Has Little Relation to Patients' Probable Benefit.* Press Release, 2012
24. Butany, J. "Coronary Artery Stents: Identification and Evaluation." *Journal of Clinical Pathology* 58.8 (2005): 795–804.
25. Newsome, Lisa T., MD, Michael A. Kutcher, MD, Sanjay K. Ghandi, MD, Richard C. Prielipp, MD, and Roger L. Royster, MD. "A Protocol for the Perioperative Management of Patients With Intracoronary Drug-Eluting Stents." *History of Coronary Stents.* APSF Committee on Education, Winter 2007. www.apsf.org/newsletters/html/2007/winter/12_protocol.htm.
26. Iqbal, J., J. Gunn, and P. W. Serruys. "Coronary Stents: Historical Development, Current Status and Future Directions." *British Medical Bulletin* 106.1 (2013): 193–211.

27. Gill, Louise Buchanan, Sandeep Basavarajaiah, and Alaide Chieffo. "Review Article: Stent Thrombosis: Incidence, Predictors and New Technologies." *Thrombosis* (2012).
28. Mehta, Rajendra H., Kishore J. Harjai, Judy A. Boura, James E. Tcheng, Simon R. Dixon, Gregg W. Stone, and Cindy L. Grines. "Short-term Outcomes of Balloon Angioplasty versus Stent Placement for Patients Undergoing Primary Percutaneous Coronary Intervention: Implications for Patients Requiring Early Coronary Artery Bypass Surgery." *American Heart Journal* 165.6 (2013): 1000–007.
29. Butany, J. "Coronary Artery Stents: Identification and Evaluation." *Journal of Clinical Pathology* 58.8 (2005): 795–804.
30. Suryapranata, H. "Long Term Outcome and Cost-effectiveness of Stenting versus Balloon Angioplasty for Acute Myocardial Infarction." *Heart* 85.6 (2001): 667–71.
31. Mitra, A. K. "In Stent Restenosis: Bane of the Stent Era." *Journal of Clinical Pathology* 59.3 (2006): 232–39.
32. Indolfi, C. "Molecular Mechanisms of In-Stent Restenosis and Approach to Therapy with Eluting Stents." *Trends in Cardiovascular Medicine* 13.4 (2003): 142–48.
33. Chatterjee, Subroto, and Ambarish Pandey. "Drug Eluting Stents: Friend or Foe? A Review of Cellular Mechanisms Behind the Effects of Paclitaxel and Sirolimus Eluting Stents." *Current Drug Metabolism* 9.6 (2008): 554–66.
34. Anderson, H. Vernon. "Drug-Eluting Stents." *Journal of the American College of Cardiology* 51.21 (2008): 2025–027.
35. Navarese, E. P., K. Tandjung, B. Claessen, F. Andreotti, M. Kowalewski, D. E. Kandzari, D. J. Kereiakes, et al. "Safety and Efficacy Outcomes of First and Second Generation Durable Polymer Drug Eluting Stents and Biodegradable Polymer Biolimus Eluting Stents in Clinical Practice: Comprehensive Network Meta-analysis." *Bmj* 347. Nov. 06 2 (2013): F6530.
36. Chitkara, Kamal, and Anthony Girshlick. "Second- versus First-generation Drug-eluting Stents." *Interventional Cardiology Review* 5.1 (2010): 23–26.
37. Machado, Carina, Luís Raposo, Hélder Dores, Sílvio Leal, Rui Campante Teles, Pedro De Araújo Gonçalves, Henrique Mesquita Gabriel, Manuel Almeida, and Miguel Mendes. "Second-generation versus First-generation Drug-eluting Stents for the Treatment of Patients with Acute Coronary Syndromes and Obstructive Coronary Artery Disease." *Coronary Artery Disease* (2014): 1.
38. Otsuka, Fumiyuki, Masataka Nakano, Elena Ladich, Frank D. Kolodgie, and Renu Virmani. "Pathologic Etiologies of Late and Very Late Stent Thrombosis following First-Generation Drug-Eluting Stent Placement." *Thrombosis* 2012 (2012): 1–16.
39. Austin, D., J. P. Pell, and K. G. Oldroyd. "Drug-eluting Stents: Do the Risks Really Outweigh the Benefits?" *Heart* 94.2 (2008): 127–28.
40. "What Is Heart Valve Disease?" *NHLBI, NIH*. US Department of Health and Human Services, 2011.
41. Gourlay, Terence, and Richard A. Black. *Biomaterials and Devices for the Circulatory System*. Oxford: Woodhead Pub., 2010.

42. Bloomfield, P. "Choice of Heart Valve Prosthesis." *Heart* 87.6 (2002): 583–89.
43. Singhal, Pooja, Adriana Luk, and Jagdish Butany. "Bioprosthetic Heart Valves: Impact of Implantation on Biomaterials." *ISRN Biomaterials* 2013 (2013): 1–14.
44. Ratner, Buddy D., Allan S. Hoffman, Frederick J. Schoen, and Jack E. Lemons. *Biomaterials Science: An Introduction to Materials in Medicine*. Burlington, MA: Elsevier Science, 2012.
45. Klabunde, Richard E. *Cardiovascular Physiology Concepts*. Philadelphia: Lippincott Williams & Wilkins/Wolters Kluwer, 2012.
46. Timms, Daniel. "A Review of Clinical Ventricular Assist Devices." *Medical Engineering & Physics* 33.9 (2011): 1041–047.
47. Barnard, James, and Steven S. L. Tsui. "The Total Artificial Heart in a Cardiac Replacement Therapy Programme." Review. *British Journal of Hospital Medicine* Dec. 2012: 634–39.
48. *Affairs of the Heart*. Perf. Barney Clarke. PBS *Scientific American*, 2001. Documentary.
49. Frazier, O. H. "First Use of an Untethered, Vented Electric Left Ventricular Assist Device for Long-term Support [Published Erratum Appears in Circulation 1995 Jun 15; 91(12):3026]." *Circulation* 89.6 (1994): 2908–914.
50. Wieselthaler, G. M., H. Schima, M. Hiesmayr, R. Pacher, G. Laufer, G. P. Noon, M. Debakey, and E. Wolner. "First Clinical Experience with the DeBakey VAD Continuous-Axial-Flow Pump for Bridge to Transplantation." *Circulation* 101.4 (2000): 356–59.
51. Altman, Lawrence K. "Implantable Heart Device Receives FDA Approval." News. *New York Times*, 6 Sept. 2006.
52. "Organ Donation and Transplantation Statistics." *National Kidney Foundation: News*. National Kidney Foundation, May 2014.
53. "The Kidneys and How They Work." *The Kidneys and How They Work Page*. National Kidney and Urologic Diseases Information Clearinghouse (NKUDIC), 21 May 2014.
54. Sullivan, John D. "End Stage Renal Disease Economics and the Balance of Treatment Modalities." *Journal of Service Science and Management* 03.01 (2010): 45–50.
55. Lewis, Adrienne. "How Hemodialysis Is Used to Treat Kidney Failure." *USA Today*, 24 Aug. 2009. Health and Behavior sec.
56. "Vascular Access for Hemodialysis." *Vascular Access for Hemodialysis*. National Kidney and Urologic Diseases Information Clearinghouse (NKUDIC), 23 July 2014.
57. Fleming, Geoffrey M. "Renal Replacement Therapy Review: Past, Present and Future." *Organogenesis* 7.1 (2011): 2–12.
58. Kerr, Peter G., and Louis Huang. "Review: Membranes for Haemodialysis." *Nephrology* 15.4 (2010): 381–85.
59. Santoro, A., and G. Guadagni. "Dialysis Membrane: From Convection to Adsorption." *NDT Plus* 3.Supplement 1 (2010): I36–39.
60. "Hemodialysis Dose and Adequacy." *Hemodialysis Dose and Adequacy*. National Kidney and Urologic Diseases Information Clearinghouse (NKUDIC), 23 Mar. 2012.
61. Bargman, Joanne M. "Advances in Peritoneal Dialysis: A Review." *Seminars in Dialysis* 25.5 (2012): 545–49.

62. Woodrow, Graham, MD, and Simon Davies, PhD. "Peritoneal Dialysis in CKD." *Peritoneal Dialysis in CKD*. The Renal Association, 30 July 2010.
63. Li, P. K.-T., C. C. Szeto, B. Piraino, J. Bernardini, A. E. Figueiredo, A. Gupta, D. W. Johnson, E. J. Kuijper, W.-C. Lye, W. Salzer, F. Schaefer, and D. G. Struijk. "Peritoneal Dialysis-Related Infections Recommendations: 2010 Update." *Peritoneal Dialysis International* 30.4 (2010): 393–423.
64. Board, A.D.A.M. Editorial. "Arthritis." *Arthritis*. U.S. National Library of Medicine, 22 Jan. 2014.
65. "Arthritis Related Statistics." Centers for Disease Control and Prevention, 2014.
66. "The Cost of Arthritis and Related Conditions." *American Academy of Orthopaedic Surgeons*. AAOS Now, Mar. 2009.
67. "Arthritis Care: Treatment for Arthritis Conditions & Arthritis Treatment Info." *Arthritis Care: Treatment for Arthritis Conditions & Arthritis Treatment Info*. The Arthritis Foundation, 2014.
68. Potts, A., J. J. Harrast, C. D. Harner, A. Miniaci, and M. H. Jones. "Practice Patterns for Arthroscopy of Osteoarthritis of the Knee in the United States." *American Journal of Sports Medicine* 40.6 (2012): 1247–251.
69. "North America Arthroscopy Procedures Outlook to 2020: Shoulder, Knee, Hip & Small Joints Arthroscopy Procedures." *Health & Medicine Week*. N.p., 18 Apr. 2014.
70. Sonntag, R., J. Reinders, and J. P. Kretzer. "What's Next? Alternative Materials for Articulation in Total Joint Replacement." *Acta Biomaterialia* 8.7 (2012): 2434–441.
71. Weinstein, Alexander M., Benjamin N. Rome, William M. Reichmann, Jamie E. Collins, Sara A. Burbine, Thomas S. Thornhill, John Wright, Jeffrey N. Katz, and Elena Losina. "Estimating the Burden of Total Knee Replacement in the United States." *Journal of Bone and Joint Surgery (American)* 95.5 (2013): 385.
72. Levine, Beth. "8 Key Facts About Knee Replacement Surgery: Will These New Knees Hold Up?" *AARP Bulletin* (n.d.): 1 Mar. 2011.
73. Windsor, Russell E., MD, and Douglas E. Padgett, MD. "Understanding Implants in Knee and Hip Replacement." *Hospital for Special Surgery*. Hospital for Special Surgery, 13 Aug. 2013.
74. CDC/National Center for Health Statistics. "In Patient Surgery." *Centers for Disease Control and Prevention*. Centers for Disease Control and Prevention, 14 May 2014.
75. Rosenthal, Elisabeth. "In Need of a New Hip, but Priced Out of the U.S." *New York Times*, 03 Aug. 2013.
76. Wheeless III, Clifford R., MD. "THR: Osteolysis." *Wheeless' Textbook of Orthopaedics*. Duke Orthopaedics.
77. Meier, Barry. "Johnson & Johnson Said to Agree to $4 Billion Settlement Over Hip Implants." *New York Times* [New York] 12 Nov. 2013.
78. Mavrogenis, A. F., R. Dimitriou, J. Parvizi, and G. C. Babis. "Biology of Implant Osseointegration." *J Musculoskelet Neuronal Interact* 9.2 (2009): 61–71.

79. Kirschner, Stephan, Jörg Lützner, Klaus-Peter Günther, Maria Eberlein-Gonska, and Frank Krummenauer. "Adverse Events in Total Knee Arthroplasty: Results of a Physician Independent Survey in 260 Patients." *Patient Safety in Surgery* 4.1 (2010): 12.
80. Revell, Peter A. "The Combined Role of Wear Particles, Macrophages and Lymphocytes in the Loosening of Total Joint Prostheses." *Journal of the Royal Society Interface* 5.28 (2008): 1263–278.
81. "Office of Research & Development." *Prosthetics/Limb Loss*. US Department of Veterans Affairs, 23 July 2014.
82. Bardhi, Ariola. "Past, Present and Future of Artificial Limb Design." Weblog post. *EJBM Blog*. WordPress.com, 11 Feb. 2013.
83. Edelstein, Joan E., and Alex Moroz. *Lower-limb Prosthetics and Orthotics: Clinical Concepts*. Thorofare, NJ: SLACK, 2011.
84. Eveleth, Rose. "Should Oscar Pistorius's Prosthetic Legs Disqualify Him from the Olympics? Scientists Debate Whether Prosthetic Legs Give Pistorius an Unfair Advantage in the 400-meter Race." *Scientific American* (2012).
85. Gailey, Robert. "Review of Secondary Physical Conditions Associated with Lower-limb Amputation and Long-term Prosthesis Use." *Journal of Rehabilitation Research and Development* 45.1 (2008): 15–30.
86. University of Southampton. "Intelligent Prosthetic Liners Could Ease Pain for Lower Limb Amputees." *Science Daily* (2014).
87. Blough, David K., Sharon Hubbard, Lynne V. Mcfarland, Douglas G. Smith, Jeffrey M. Gambel, and Gayle E. Reiber. "Prosthetic Cost Projections for Servicemembers with Major Limb Loss from Vietnam and OIF/OEF." *Journal of Rehabilitation Research and Development* 47.4 (2010): 387.
88. Fougner, A., O. Stavdahl, P. J. Kyberd, Y. G. Losier, and P. A. Parker. "Control of Upper Limb Prostheses: Terminology and Proportional Myoelectric Control—A Review." *IEEE Transactions on Neural Systems and Rehabilitation Engineering* 20.5 (2012): 663–77.
89. Billock, John N., CPO. "Upper Limb Prosthetic Terminal Devices: Hands Versus Hooks." *Clinical Prosthetics and Orthotics* 10.2 (1996): 57–65.
90. Swiontkowski, M. F. "Targeted Muscle Reinnervation for Real-time Myoelectric Control of Multifunction Artificial Arms." *Yearbook of Orthopedics* 2010 (2010): 91–93.
91. Young, Aaron J., Lauren H. Smith, Elliott J. Rouse, and Levi J. Hargrove. "A Comparison of the Real-time Controllability of Pattern Recognition to Conventional Myoelectric Control for Discrete and Simultaneous Movements." *Journal of NeuroEngineering and Rehabilitation* 11.1 (2014): 5.
92. Manson, Theodore T., MD, MS, and Andrew J. Cosgarea, MD. "Meniscal Injury in Active Patients." *Advanced Studies in Medicine* 4.10 (2004): 545–52.
93. Di Matteo, Berardo, Vittorio Tarabella, Giuseppe Filardo, Anna Viganò, Patrizia Tomba, and Maurilio Marcacci. "Thomas Annandale: The First Meniscus Repair." *Knee Surgery, Sports Traumatology, Arthroscopy* 21.9 (2013): 1963–966.

94. Laible, C., D. A. Stein, and D. N. Kiridly. "Meniscal Repair." *Journal of the American Academy of Orthopaedic Surgeons* 21.4 (2013): 204–13.
95. Provencher, Matthew T., MD, and Daniel J. Solomon, MD. "Meniscal Allografts." *Surgical Management of Articular Cartilage Defects in the Knee*. Ed. Matthew W. Rose. N.p.: American Academy of Orthopaedic Surgeons, 1–16.
96. Crook, T. B., A. Ardolino, L. A. P. Williams, and I. W. Barlow. "Meniscal Allograft Transplantation: A Review of the Current Literature." *Annals of the Royal College of Surgeons of England* 91.5 (2009): 361–65.
97. Jiang, Dong, Li-Heng Zhao, Ming Tian, Ji-Ying Zhang, and Jia-Kuo Yu. "Meniscus Transplantation Using Treated Xenogeneic Meniscal Tissue: Viability and Chondroprotection Study in Rabbits." *Arthroscopy: Journal of Arthroscopic & Related Surgery* 28.8 (2012): 1147–159.
98. Humes, H. David, Deborah Buffington, Angela J. Westover, Shuvo Roy, and William H. Fissell. "The Bioartificial Kidney: Current Status and Future Promise." *Pediatr Nephrol Pediatric Nephrology* 29.3 (2013): 343–51. Web.
99. Pino, Christopher J., and H. David Humes. "Stem Cell Technology for the Treatment of Acute and Chronic Renal Failure." *Translational Research* 156.3 (2010): 161–68. Web.
100. Fissell, William H., Aaron J. Fleischman, H. David Humes, and Shuvo Roy. "Development of Continuous Implantable Renal Replacement: Past and Future." *Translational Research* 150.6 (2007): 327–36. Web.
101. Johnston, K., and H. D. Humes. "Extracorporeal Renal Replacement." *Principles of Regenerative Medicine*. By Anthony Atala. 2nd ed. Amsterdam: Elsevier/Academic, 2011. 943–53. Print.
102. Fissell, W. "Implanted Renal Replacement for End-stage Renal Disease." *Panminerva Med* 53.3 (2011): 155–66. Web.
103. Humes, H. David, Sherrill M. Mackay, Angela J. Funke, and Deborah A. Buffington. "Tissue Engineering of a Bioartificial Renal Tubule Assist Device: In Vitro Transport and Metabolic Characteristics." *Kidney International* 55.6 (1999): 2502–514. Web.

CHAPTER 5
Reproductive Technologies

a. In Vitro Fertilization

Over the past thirty years, women have given birth to other people's children; sperm are routinely sorted to influence the gender of the offspring; and preimplant embryos are screened for the presence or absence of genetic diseases. As of 2013, other relatively commonplace events in reproductive medicine include artificial insemination, surrogate motherhood, pregnancy in postmenopausal women, and freezing and long-term storage of human embryos. One of the most controversial developments in the field is commerce in unfertilized eggs. Egg donors can be paid between $2,000 and $100,000 to undergo a series of injected hormone treatments to induce hyperovulation, followed by diagnostic imaging for tracking the developing egg follicles and removal of the mature eggs (1). Mature eggs are then fertilized in vitro (within a petri dish in the laboratory), with sperm typically collected from the recipient couple. Following fertilization, these cells are allowed to mature to the blastocyst stage (eight- to sixteen-cell stage) (Figure 5.1), screened for genetic abnormalities, and then are implanted into the uterus of the recipient female (2, 3). Additional eggs may be cryopreserved in liquid nitrogen for fees that can range up to—and in excess

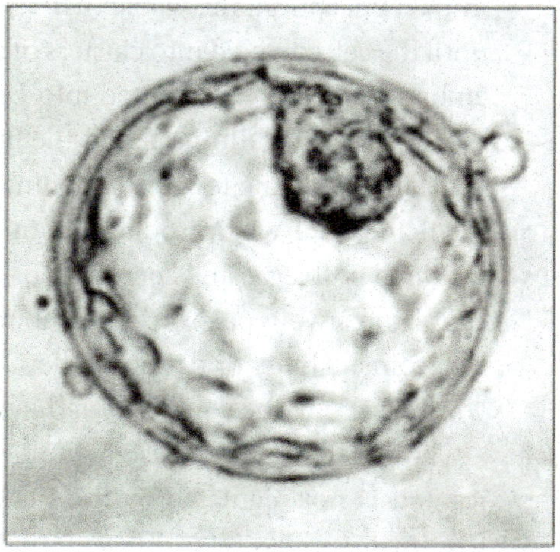

FIGURE 5.1: Human Blastocyst
Conaghan / NIH / Copyright in the Public Domain.

of—$8,000 per year (4, 5). The pursuit of egg donations from Ivy League women has increased with the placement of advertisements in college newspapers and on the Internet, and higher-end monetary compensations are afforded this specific population of donors.

Over the past one hundred years, mother and infant death rates have decreased significantly. In 1901, there were approximately ten infant deaths per one hundred live births in the United States. As of 2013, this percentage was approximately 0.6 percent (6). Mortality of the delivering mother also decreased from 1.0 percent to 0.02 percent in the United States during this same time period (6). Advances in maternal health care, infection control, and improvements in maternal nutrition have all contributed to improved outcomes in maternal and neonatal medicine. As society has progressed, many women are choosing to begin families at later ages, with approximately 15 percent to 20 percent of women giving birth for the first time at age thirty-five years and older (7). Women in this age range generally have more difficulties either becoming pregnant or carrying their pregnancies to full term. Issues related to the age of the eggs and the numbers of eggs available are only two of a wide range of possible reasons for decreases in fertility. In fact, one in six couples experience infertility when attempting to conceive (8). According to recent data published by the CDC, 10 percent of women, 6.1 million in the United States, ages fifteen to forty-four have difficulties becoming pregnant or physiologically maintaining their pregnancies during the gestational period (8). Despite major advances in medicine, some of which have already been mentioned, the overall success rate of treating idiopathic infertility that is unresponsive to pharmaceutical intervention is less than 25 percent.

The modern era of assisted reproductive technology (ART) began in 1978 with the birth of Louise Brown in the United Kingdom (9). Unfertilized eggs from Louise's mother, Lesley Brown, were harvested and fertilized in a petri dish with sperm harvested from Louise's father. Following initial conception, the viable (living) zygote (two cells) matured in vitro for approximately six days until the cells divided into clusters of eight to sixteen cells. Several blastocysts were considered viable embryos and were implanted into Lesley Brown's uterus. One embryo attached to the uterine wall and developed, culminating in the birth of Louise. In 2010, Dr. Robert G. Edwards won the Nobel Prize for Medicine for this crowning achievement. Following this landmark event in reproductive medicine, ART was introduced in the United States in 1981 (9). ART includes all fertility treatments in which both eggs and sperm are handled. In general, ART procedures involve surgically removing eggs from a woman's ovaries, combining them with sperm in the laboratory, and returning them to the woman's body or donating them to another woman (Figure 5.2) (10). Thus far, 5 million babies worldwide have been born with the aid of ART or, more specifically, in vitro fertilization (IVF) (11).

Within the United States, the IVF process is typically carried out in for-profit medical clinics and is not regulated by the FDA. Fertility clinics are required to register and report annual live-birth success rates. These reports are utilized as a resource by consumers to choose fertility clinics. Embryology laboratories, which perform the manipulations of eggs and sperm, are subject to inspections by the College of American Pathologists in compliance with scientific regulations. The average cost of IVF in the United States is $13,000 per cycle (12). A cycle involves the steps leading to implantation of the embryo in the recipient's uterus, as described above. The high price

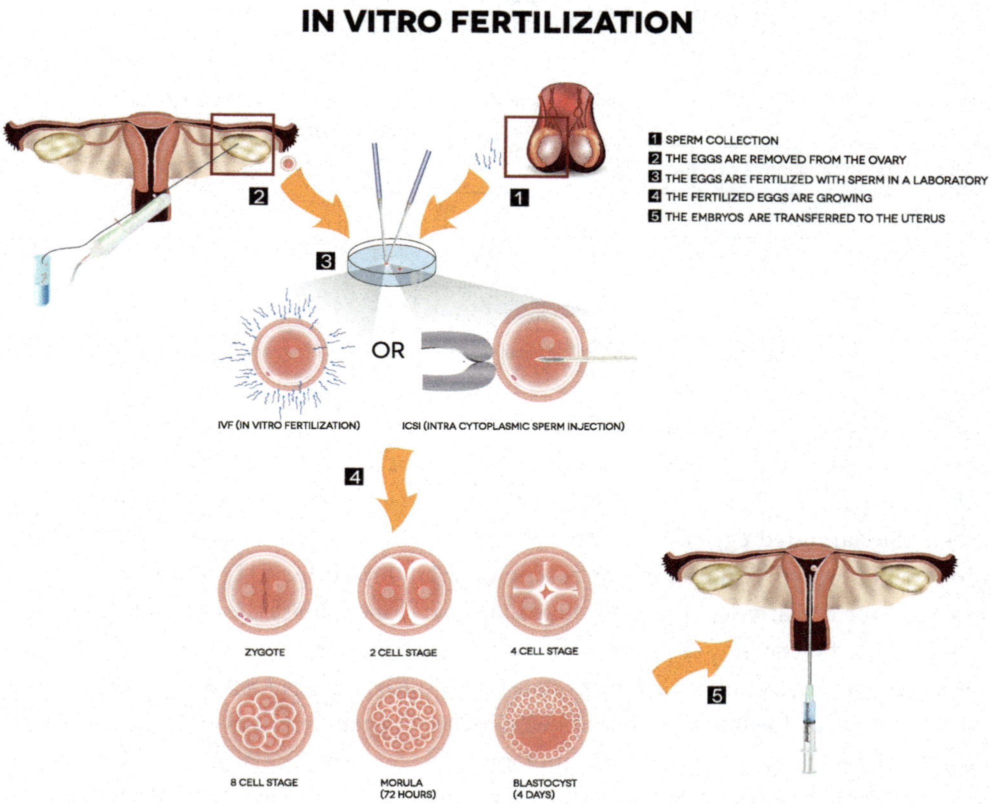

FIGURE 5.2: Assisted Reproductive Technology
Copyright © 2013 by Depositphotos / megija.

tag reflects the cost of health coverage in the United States and sharply contrasts to similar procedures in other countries such as Japan, which run $4,000 (13). Implantation of several embryos at one time increases the likelihood of live births and also increases the success rates of the specific fertility clinics and ultimately profitability.

The IVF process can be grueling for egg donors and embryo recipients alike and can include ovarian stimulation, in vitro treatment of gametes, intracytoplasmic sperm injections, cryopreservation of eggs and/or embryos, and in vitro maturation of eggs. During the first week of the patient's menstrual cycle, the patient is administered oral contraceptives in order to regulate the menstrual cycle and prime the ovaries (Figure 5.3). If a donor is providing the eggs, the donor is given oral contraceptives in order to regulate her menstrual cycle with the embryo recipient's and will abide by the following protocol. The drug Lupron® is also administered in order to prevent premature release of the developing eggs. At the beginning of what would be the next menstrual cycle, follicle stimulating hormone (FSH) is administered by the patient at home by daily injections. This is carried out for approximately twelve days, and the maturation of the follicles and development of the

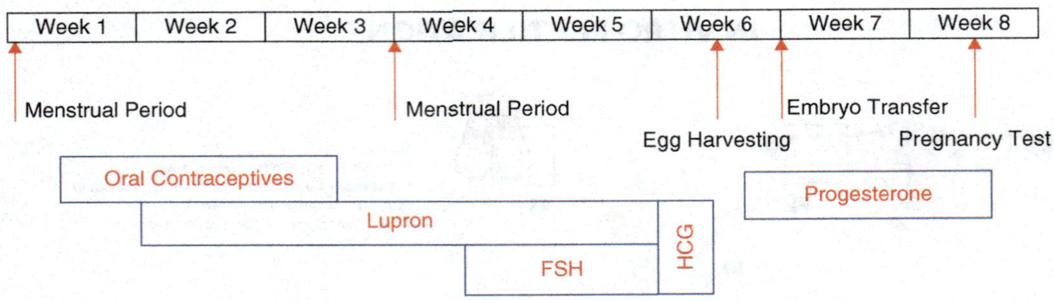

FIGURE 5.3: Assisted Reproductive Technology: Hormone Injection, Egg Harvest, and Fertilization Cycle (original figure)

eggs are monitored using ultrasound imaging. At this point, when the physician determines the maturity of the eggs, the eggs are harvested via ultrasound-guided needle aspiration under general anesthesia. The harvested eggs are fertilized with sperm in vitro and remain in culture until the formation of blastocysts is noted by microscopic evaluation. Prior to fertilization, eggs may be cryopreserved for a yearly fee. Once fertilized and developed, blastocysts—also called embryos at this stage—can be cryopreserved as well. Two to seven embryos are then implanted into the recipient's uterus using a guided needle. As stated previously, increasing the numbers of implanted embryos increases the likelihood of obtaining a viable pregnancy, and over 5 million babies have been born worldwide as a result of this technology (11). However, implantations of many embryos can lead to multiple births with resultant developmental complications in the developing infants. An extreme example is the case of Nadya Suleman, who in 1999 was implanted with all eight viable embryos and gave birth to all eight (albeit preterm) infants (14). This sparked outrage in the general public as well as the scientific community. Dr. Michael Kamrava, Ms. Suleman's physician in charge of her IVF procedures, was banned from the American Society for Reproductive Medicine on the basis that the implantation of so many embryos in a woman already caring for six children from previous IVF treatments was unethical and reckless (15).

IVF, although invaluable for many infertile couples, is extremely expensive and not covered by insurance. Many patients require several cycles in order to become pregnant, with success rates averaging 25 percent (10). In order to address this fairly low success rate, techniques have been developed to aid in fertilization and ensure viability of the embryo. Such techniques can be considered "extreme IVF."

b. Specialized Reproductive Technologies

IVF is an expensive and grueling option for many infertile couples. Advancements in fertilization and implantation procedures, manipulations of sperm and eggs, and screening of developing embryos have expanded some of the options for these couples.

Couples who have difficulties with in situ fertilization between healthy sperm and healthy eggs directly benefit from traditional IVF procedures. When a woman cannot produce eggs appropriate for successful pregnancy, eggs can be obtained from a third-party egg donor, as stated previously, and a substantial donor fee is often required. When a woman produces satisfactory eggs but cannot carry them to pregnancy, then a surrogate may be engaged to bear the child. Over ten thousand surrogate pregnancies have occurred in the United States as of 2013 (16). Surrogate mothers are implanted with the embryos belonging to the infertile couples that were conceived by IVF. Surrogates can receive from $12,000 to $25,000 per pregnancy, and they must waive their rights to legal guardianship of the children. Controversies exist regarding the legal rights of surrogates as well as those of the biological parents as they pertain to compensations and legal connections to the children. Finally, IVF can be used to provide children to same-sex partnerships, though each offspring will carry the genes of one parent and one donor.

Advancements in in vitro manipulation techniques of sperm and eggs have optimized fertilizations when physiological conditions have hindered natural processes (Table 5.1). Sperm may be directly injected into eggs, which is most useful in males with very low sperm counts. This process is called intracytoplasmic sperm injection (ICSI) (17). Assisted hatching, opening the zona pellucida (embryo shell), is a technique that aids in the hatching of embryos (18). This increases the chances of successful implantation into the lining of the uterus when, under normal physiological conditions, viable embryos would not be capable of implanting on their own. Removal of cellular fragments that may hinder development of the embryo is also possible using micropipette manipulations in the petri dish. Two controversial manipulations involve removal of cytoplasm from an unfertilized donor egg, followed by injection of the cytoplasm into the patient's egg along with the sperm (19). This provides a physiologically optimal environment for fertilization in situations where the patient's egg cannot support fertilization. The second controversial manipulation is removal of the pronuclei from a fertilized egg followed by its transfer into an enucleated egg donated

TABLE 5.1: Extreme Assisted Reproductive Technologies (original table)

Technique	Description
Intracytoplasmic sperm injection	Direct injection of donor sperm into donor egg in vitro
Assisted hatching	Opening of the egg's zona pellucida in order to facilitate implantation into the uterine lining
Cellular fragmentation	Removal of cellular debris from the cytoplasm of the blastocyst in order to enhance viability
Cytoplasmic transfer	Removal of cytoplasm from an unfertilized donor egg followed by injection of the cytoplasm into the patient's egg along with the sperm
Pronuclei transfer	Removal of the pronuclei from a fertilized egg followed by its transfer into an enucleated egg donated by another
Gender selection	Selection of either male or female sperm based upon differences in density

by another (20). The reconstructed embryo is transferred back to the biological mother. This method is potentially useful in women whose eggs lack the cytoplasm to create viable blastocysts. Both procedures have the potential to mix mitochondrial DNA from the egg donor with nuclear DNA from the biological mother. The FDA currently has a hold on the use of this technique, despite its broader policy of not regulating the practice of reproductive medicine. Cytoplasmic transfer is legal in some countries, and US couples have gone offshore for the therapy.

In the simplest case of influencing the characteristics of offspring, sperm is processed through a "cell sorter" that separates sperm on the basis of the slight difference in weight of the larger X and smaller Y chromosomes (21). The process costs $3,200 and is 91 percent successful in producing female offspring and 76 percent successful in producing male offspring. This process is not widely used in the United States.

Finally, the development of preimplantation genetic diagnosis (PGD) now allows prospective parents to select against embryos displaying genetic and chromosomal abnormalities prior to implantation (22). Extra copies of a chromosome (trisomy) or missing copies of a chromosome (monosomy) often leads to miscarriage. In this context, PGD has a dual purpose: to increase the likelihood of successful pregnancy, as well as to eliminate the disease in a child if an aberrant chromosome were carried to term. In addition, blastocysts can be tested for over a hundred single-gene disorders, including cystic fibrosis, fragile X syndrome, and Tay-Sachs disease. When the developing embryo has reached the eight-cell stage, at approximately three days postfertilization, one to two cells are removed and biopsied. Specific segments of DNA of the harvested cells are amplified using polymerase chain reaction (PCR) and then tested for the presence of monogenic diseases using molecular diagnostic techniques such as automated DNA nucleotide sequencing. Fluorescence labeling of chromosomes, fluorescence in situ hybridization (FISH) (Figure 5.4), is used for the determination of X-linked diseases, chromosomal abnormalities, and aneuploidy screening (23). PGD is essentially equivalent to amniocentesis, except that it avoids the risks associated with that procedure, eliminates the need for abortion, and may avoid implantation of a blastocyst that is likely to miscarry.

Technology on the Edge: Mitochondrial Replacement in In Vitro Fertilization

The human mitochondria contain approximately sixty thousand base pairs that encode for thirty-seven genes that are maternally inherited. Mutations in any of these genes or nuclear genes that control mitochrondrial functions can result in debilitating metabolic disorders affecting the heart, brain, and muscle. Tissues with high metabolic demand are most often affected. Approximately one in four thousand children in the United States per year are born with mitochondrial defects resulting in severe disabilities (24). Severity of disease varies, and patients' life expectancies are shortened due to associated conditions such as diabetes and heart, kidney, and liver failure.

Since no curative treatments exist for mitochondrial diseases, scientists and physicians are focusing on options made possible by in vitro fertilization and preimplantation genetic diagnosis.

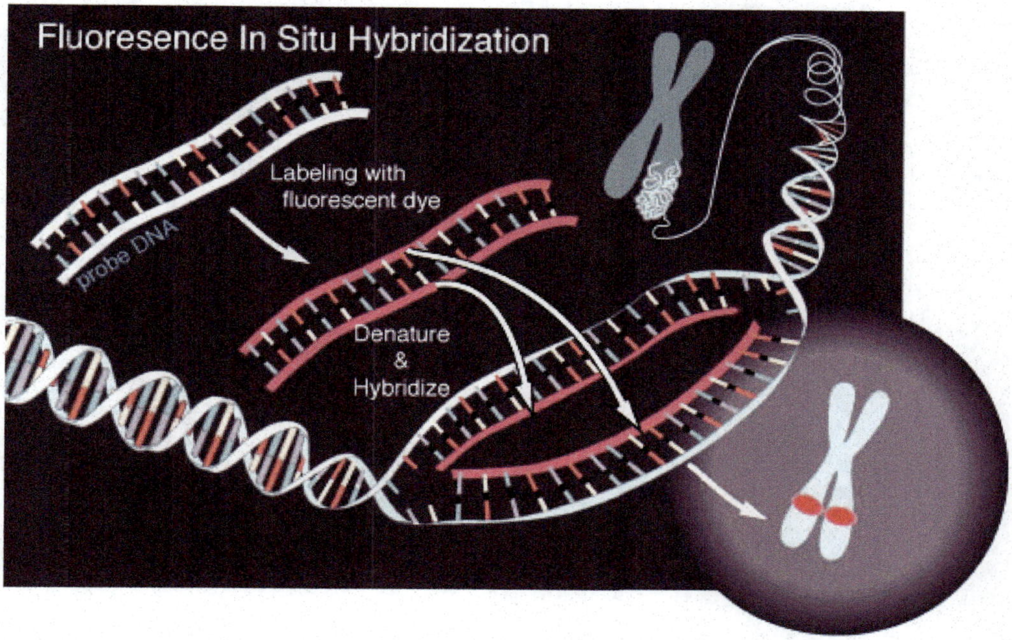

FIGURE 5.4: Fluorescent In Situ Hybridization
Thomas Ried / National Human Research Genome Institute / Copyright in the Public Domain.

Three potential methods for preventing transfer of defective mitochondrial DNA are cytoplasmic transfer, pronuclear microinjection, and spindle transfer.

Cytoplasmic transfer during the process of in vitro fertilization was first performed in humans in 1997 (25). This process involves transfer of a small volume of ooplasm from a donor egg into the ooplasm of the recipient's affected egg. This method does not remove the defective mitochondrial DNA from the recipient's egg but does introduce a normal mitochondrial genome to support overall cellular function. The resulting embryo carries the genome of three lineages: donor egg mitochondrial DNA, recipient egg mitochondrial DNA and nuclear DNA, and sperm nuclear DNA.

An alternative to cytoplasmic transfer is pronuclear transfer. Prior to fertilization, pronuclei from the egg and sperm are removed and both are transferred into the ooplasm of a donated, enucleated egg. Following electrical stimulation, fertilization occurs, and theoretically normal embryonic development ensues. This procedure has been achieved in mouse models as well as human zygote models, albeit with varying degrees of success (26, 27).

Spindle transfer provides a third option for treating mitochondrial diseases, and the technique parallels that used for somatic cell nuclear transfer. In this procedure, instead of transferring the nucleus of a somatic cell into an enucleated egg, the spindle of the affected egg, along with the attached maternal chromosomes, are transferred to a donor enucleated egg (28). The egg is then fertilized according to standard in vitro fertilization protocols. Thus far, success has been demonstrated in primate models with the birth of healthy offspring (28).

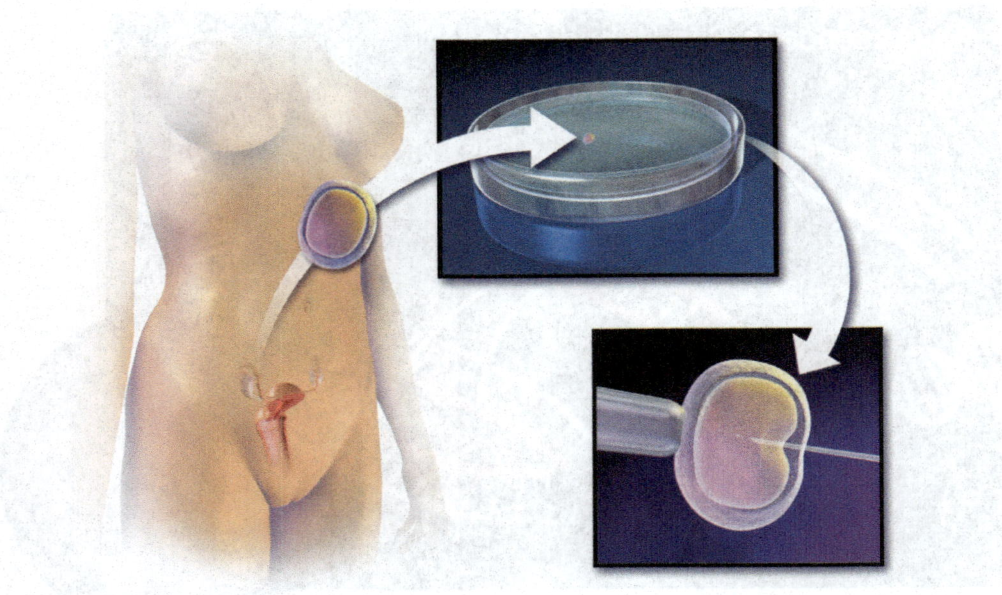

FIGURE 5.5: In Vitro Manipulation of Unfertilized Egg
Copyright © 2014 by BruceBlaus / Wikimedia Commons, (CC BY 3.0) at https://commons.wikimedia.org/wiki/File:Blausen_0583_InVitroFertilization.png.

The three in vitro fertilization procedures stated above are becoming viable options for patients who carry inherited defects in their mitochondrial genome. Complete success in the laboratory has not been achieved, however, as carryover of mutated mitochondrial DNA has been shown. Optimization of microsurgery and overall in vitro transfer procedures are necessary. Additionally, ethical issues continue to surround genetic manipulation of egg, sperm, and embryo.

End-of-Chapter Questions and Answers

1. What is the overall success rate of treating idiopathic infertility that is unresponsive to pharmaceutical intervention?
   25 percent
2. What is the pay range for women who donate their eggs for in vitro fertilization?
   $2,000–$100,000
3. Women intending to donate their eggs are placed on hormone therapy. What physiological response does this therapy induce?
   Hyperovulation
4. Once eggs are harvested, they are fertilized by sperm in vitro and are allowed to mature to the eight- to sixteen-cell stage. What is the name for this cell cluster?
   Blastocyst

5. What type of genetic analysis is typically performed on the eight- to sixteen-cell cluster prior to implantation?

   Preimplantation genetic diagnosis

6. Several novel technologies have been developed in order to enhance fertilization of the egg prior to implantation. Which technique involves injections of sperm directly into the cytoplasm of the mature egg?

   Intracyotplasmic sperm

7. Opening of the zona pellucida (embryo shell) is a technique that aids in the hatching of embryos. What is this technique called?

   Assisted hatching

8. Which technique is used for determining X-linked diseases, chromosomal abnormalities, and aneuploidy in embryos considered for implantation?

   Fluorescence labeling of chromosomes, fluorescence in situ hybridization (FISH)

9. Which organelle is transferred during the removal of cytoplasm from an unfertilized donor egg followed by injection of the cytoplasm into the patient's egg?

   Mitochondria

10. Which method is potentially useful in women whose eggs lack the cytoplasm to create viable blastocysts?

   Pronuclear transfer

Works Cited

1. Rabin, Roni Carin. "As Demand for Donor Eggs Soars, High Prices Stir Ethical Concerns." *New York Times*, 15 May 2007.
2. Wang, Jeff, and Mark V. Sauer. "In Vitro Fertilization (IVF): A Review of 3 Decades of Clinical Innovation and Technological Advancement." *Therapeutics and Clinical Risk Management* 2.4 (2006): 355–64.
3. "In Vitro Fertilization Information—The Mount Sinai Hospital." *Mount Sinai Hospital*. New York: Mount Sinai, 2014.
4. Saragusty, J., and A. Arav. "Current Progress in Oocyte and Embryo Cryopreservation by Slow Freezing and Vitrification." *Reproduction* 141.1 (2010): 1–19.
5. Robertson, John A. "Egg Freezing and Egg Banking: Empowerment and Alienation in Assisted Reproduction." *Journal of Law and the Biosciences* (2014).
6. "MMWR Weekly: Current Volume (2014)." *Centers for Disease Control and Prevention*. Centers for Disease Control and Prevention.
7. Martinez, Gladys, PhD, Kimberly Daniels, PhD, and Anjani Chandra, PhD. *Fertility of Men and Women Aged 15–44 Years in the United States: National Survey of Family Growth, 2006–2010*. Rep.: Centers for Disease Control and Prevention, *Centers for Disease Control and Prevention*. Centers for Disease Control and Prevention.

8. "Infertility FAQs." *Centers for Disease Control and Prevention*. Centers for Disease Control and Prevention, 20 June 2013.
9. Kamel, Remah Ma. "Assisted Reproductive Technology after the Birth of Louise Brown." *Gynecology & Obstetrics* 03.03 (2013).
10. Brown, Rachel, and Joyce Harper. "The Clinical Benefit and Safety of Current and Future Assisted Reproductive Technology." *Reproductive BioMedicine Online* 25.2 (2012): 108–17.
11. Loendersloot, Laura Van, S. Repping, P. M. M. Bossuyt, F. Van Der Veen, and M. Van Wely. "Prediction Models in in Vitro Fertilization; Where Are We?" *Journal of Advanced Research* (2013).
12. Brown, Haywood L., MD. "Preconceptional Considerations and Counseling for the Infertile Couple." *Ncmedicaljournal* 70.5 (2009).
13. Fauser, B. C. J. M. "Alternative Approaches in IVF." *Human Reproduction Update* 8.1 (2002): 1–9.
14. Bowe, John. "The Octomom and Her Babies Prepare for Prime Time." *New York Times*, 12 Nov. 2009.
15. Mohajer, Shaya Tayefe. "Octomom's Doctor Has License Revoked For 'Mega–Birth.'" *Huffington Post*, 1 Aug. 2011.
16. Cohen, Deborah L. "Surrogate Pregnancies on Rise despite Cost Hurdles." *Reuters* [Chicago], 18 Mar. 2013.
17. Palermo, Gianpiero, MD. "Intracytoplasmic Sperm Injection." *Practical Manual of In Vitro Fertilization*. New York: Springer, 2012.
18. Hammadeh, Mohamad Eid, Constanze Fischer-Hammadeh, and Khaled Refaat Ali. "Assisted Hatching in Assisted Reproduction: A State of the Art." *Journal of Assisted Reproduction and Genetics* 28.2 (2011): 119–28.
19. Nagy, Zsolt Peter, Alex C. Varghese, and Ashok Agarwal. "Nuclear and Cytoplasmic Transfer: Human Applications and Concerns." *Practical Manual of in Vitro Fertilization: Advanced Methods and Novel Devices*. New York: Springer, 2012. 659–65.
20. Craven, Lyndsey, Helen A. Tuppen, Gareth D. Greggains, Stephen J. Harbottle, Julie L. Murphy, Lynsey M. Cree, Alison P. Murdoch, et al. "Pronuclear Transfer in Human Embryos to Prevent Transmission of Mitochondrial DNA Disease." *Nature* 465.7294 (2010): 82–85.
21. Schulman, Joseph D., and David S. Karabinus. "Scientific Aspects of Preconception Gender Selection." *Reproductive BioMedicine Online* 10 (2005): 111–15.
22. Harper, Joyce C., and Sioban B. SenGupta. "Preimplantation Genetic Diagnosis: State of the ART 2011." *Human Genetics* 131.2 (2012): 175–86.
23. Gardner, David K., Denny Sakkas, Emre Seli, and Dagan Wells. "Preimplantation Genetic Screening and Diagnosis Using Fluorescent In Situ Hybridization (FISH)." *Human Gametes and Preimplantation Embryos: Assessment and Diagnosis*. New York: Springer, 2013. 125–34.
24. Schaefer, A. M., and P. F. Chinnery. "The Epidemiology of Mitochondrial Disorders: Past, Present and Future." *Biochim Biophys Acta*. 1659 (n.d.): 115–20. Web.

25. Cohen, Jacques, Richard Scott, Tim Schimmel, Jacob Levron, and Steen Willadsen. "Birth of Infant after Transfer of Anucleate Donor Oocyte Cytoplasm into Recipient Eggs." *Lancet* 350.9072 (1997): 186–87. Web.
26. McGrath, J., and D. Solter. "Nuclear Transplantation in the Mouse Embryo by Microsurgery and Cell Fusion." *Science* 220.4603 (1983): 1300–302. Web.
27. Craven, Lyndsey, Helen A. Tuppen, Gareth D. Greggains, Stephen J. Harbottle, Julie L. Murphy, Lynsey M. Cree, Alison P. Murdoch, et al. "Pronuclear Transfer in Human Embryos to Prevent Transmission of Mitochondrial DNA Disease." *Nature* 465.7294 (2010): 82–85. Web.
28. Tachibana, Masahito, Michelle Sparman, Hathaitip Sritanaudomchai, Hong Ma, Lisa Clepper, Joy Woodward, Ying Li, et al. "Mitochondrial Gene Replacement in Primate Offspring and Embryonic Stem Cells." *Nature* 461.7262 (2009): 367–72. Web.

CHAPTER 6

Tissue and Stem Cell Engineering

a. A History of Tissue Engineering

Tissue engineering is the process of manipulating cells outside of the body with concomitant formation of novel tissue constructs and implantation of these constructs in the attempts to repair or restore damaged tissues and organs. For thousands of years, humans have strived to replace lost functions of tissues and organs by tissue substitution (1, 2). The earliest clinically successful attempt of tissue substitution was in the area of skin transplantation (2). Thus, advancements in tissue and organ transplantation provided incremental validation for the burgeoning field of tissue engineering.

As stated in the definition, tissue engineering relies on the manipulation and maintenance of cells outside of the body. These first steps could not have been possible without the development of in vitro cell culture. Fundamental biological experimentation by Rudolf Virchow from 1821–1902 established that the regeneration of tissues requires cell proliferation (2). During the late nineteenth and early twentieth centuries, critical breakthroughs in the developments of in vitro cell manipulation and cell culture maintenance provided the foundations of modern-day tissue engineering (3, 4). Surgical manipulation of tissues dominated medicine in the decades following these seminal experiments, and it was not until the 1970s that cells were combined with structural matrices in vitro in order to replicate natural tissue function (2). Investigators Yannas and Burke from MIT revolutionized the field in the 1980s by producing functional 2-D skin in vitro by growing dermal fibroblasts on protein scaffolds. It was then that the term "tissue engineering" was coined and one of the first biotechnology companies, Organogenesis, Inc. was established (2). Drs. Joseph Vacanti from Boston Children's Hospital and Robert Langer from MIT were also integral in presenting definitive scientific evidence that validated the basic principles and practices of modern-day tissue

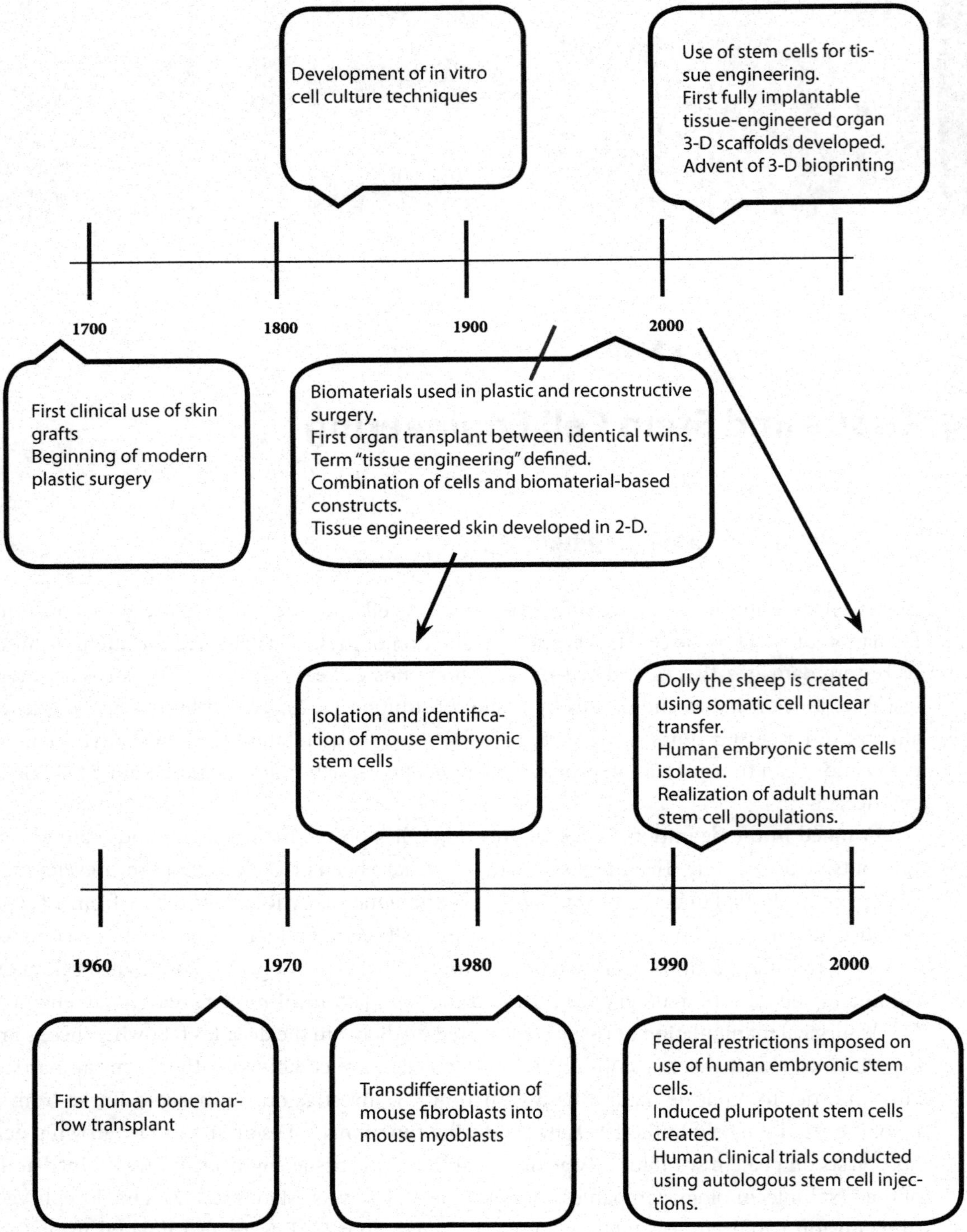

FIGURE 6.1: Generalized Time Line of Tissue Engineering and Stem Cell Technologies (original schematic)

engineering (5). Following these accomplishments in the laboratory, this fairly new discipline of tissue engineering was thrust into the public arena by Dr. Charles Vacanti from the University of Massachusetts Medical Center, who published reports showing a tissue-engineered ear grown on the back of a mouse (2). Experimentation by laboratories worldwide has grown exponentially in the twentieth and twenty-first centuries with investigators using a variety of cell and tissue types, as well as biological, synthetic, and combination scaffolds and hundreds of types of growth factors. The most recent innovation has been the design of tissues in vitro using 3-D biological printing technology that was modeled on cartridge printers for 2-D printing applications.

Regenerative medicine, the use of cells to repair and replace tissue function in situ, is now being used clinically, albeit in limited cases. A general history of stem cells, the primary cells employed for in situ repair, is discussed below (Figure 6.1).

b. A History of Stem Cells

Stem cells constitute the earliest cells that lay the foundation of all human tissues. The result of fertilization of a female egg by a male sperm is the formation of a diploid zygote that continues to evenly divide into clusters of 8, 16, and 32 cells called a blastocyst. The cells forming the blastocyst form the embryo and are called embryonic stem cells. These cells have the abilities to divide indefinitely and to differentiate, or become, any cell type and hence any tissue type in the human body. These cells considered are pluripotent. Upon further mitotic cycles, the blastocyst differentiates—according to physical, electrical, and biochemical cues—into the complex tissues of the fetus. During fetal development, cells continue to mature, thus forming the intricacies of human organ systems. Birth hails changes in cellular development with continued maturation of tissues and sequestration of niches of stem cells, albeit with decreased mitotic capacities and potencies. These cells are considered unipotent.

In the quest to regenerate failing tissues and organs, investigators have searched for early cell types such as stem cells to provide the building blocks for tissue reconstruction. Several cell types have been discovered—embryonic, adult, and induced pluripotent—that have the capacity to structurally rebuild damaged adult tissues. These cells will be described in detail later in the chapter.

Stem cell biology has been progressing rapidly for over thirty-five years. The first use of stem-like cells to replenish human tissues was in 1968 with the implementation of the first bone marrow transplant (6). Although knowledge existed regarding bone marrow's ability to replenish depleted blood cell lines in immunosuppressed patients, the acknowledgement of these cells as stem-like cells did not occur for another twenty years. The discovery of the first mammalian embryonic stem cells was made in a mouse model by Martin Evans while working at the University of Cambridge in the United Kingdom in 1981 (7). Following this critical discovery, investigators raced to isolate embryonic stem cells from the blastocysts of other species, including human. In 1998, James Thomson of the University of Wisconsin in Madison and John Gearhart of Johns Hopkins University in

Baltimore, respectively, isolated human embryonic stem cells in their laboratories (8). A year prior, Ian Wilmut and his colleagues at the Roslin Institute at the University of Edinburgh, Scotland, created the technique of somatic cell nuclear transfer. An egg from a sheep was harvested, enucleated and, through electrical stimulation, was fused with the nuclear DNA of a mature udder cell from a donor animal. The investigators proceeded to grow the newly fertilized egg in the laboratory until it reached the blastocyst stage and then transferred the embryo into the uterus of a surrogate sheep. Over two hundred attempts were required before the viable birth of Dolly the sheep in 1997 (9). Dolly was considered a cloned animal, although her nuclear and mitochondrial DNA were extracted from two different sources. Current use of somatic cell nuclear transfer is termed "therapeutic cloning," where the newly formed blastocyst is used as a cell reservoir to grow stem cells in vitro with the intent of using these cells to regenerate tissue upon implantation.

The concept of developing cloned animals excited as well as shocked the scientific community and public at large. Ethical concerns over the use of human embryos and their identities as complete human organisms spurred President George W. Bush to place limits on federal funding of research on human embryonic stem cells (10). Sixty-four human stem cell lines had been established by that point, and investigators were allowed to continue their research on these specific lines. Unfortunately, as time went by, viabilities of the human embryonic stem cell lines decreased due to the freezing and thawing cycles that were required to begin in vitro cultures. During the next four years, laboratories around the world raced to characterize the stem cells that formed the blastocyst of the early embryo in order to use them to treat diseases and disorders ranging from Parkinson's disease to traumatic spinal injury and paralysis. Even under increasing governmental limitations and public scrutiny, the desire for pluripotent embryonic stem cells and their use in tissue regeneration grew. In 2006, Shinya Yamanaka of Kyoto University in Japan developed an alternative method for creating stem cells by dedifferentiating, or transposing, adult cells into induced pluripotent stem cells (11). Yamanaka's laboratory screened dozens of oncogenes, cancer-inducing genes, in order to determine which genes were necessary to genetically switch adult cells back into a stemlike state. This technique changed the transcriptional programming of the cells, resulting in the formation of pluripotent cells. Along with the potential for these cells to be used therapeutically to regenerate tissues, these cells also carried the risk of uncontrolled cell division and mutation, much like cancer cells. Several methods have since been developed to produce pluripotent stem cells from adult somatic cells without the threat of oncogenesis.

In 2009, President Barack Obama eased the restrictions on federal funding for human embryonic stem cell research (12). Investigators' applications for federal funding of human embryonic stem cell research are now intensely screened by the National Institutes of Health prior to being considered for support, and private funding of human embryonic stem cell research is scrutinized, using slightly different parameters, by the research institution and private funding party. Since 2008, the research and medical communities have worked endlessly to characterize embryonic, adult niche, and reprogrammed stem cells, as well as to introduce novel stem cell–based therapies into the clinic. The technique of therapeutic cloning has been employed to develop insulin-producing beta cells from human stem cells originating from somatic cell nuclear transfer (13).

Treatments for traumatic spinal injury, retinal blindness, and cardiovascular disease have also been created with stem cell–based implants. According to Clinicaltrials.gov, there are currently over two thousand human clinical trials involving stem cell–based therapies in process in the United States (14).

c. Strategies for Tissue and Stem Cell Engineering

i. Cells

Successful tissue engineering constructs require, at a minimum, functional cell types, supportive matrices, and soluble or matrix-bound biochemical factors (Figure 6.2). Cells may be harvested from a variety of sources, including primary tissues, immortalized cell lines, and stem cells (Table 6.1). Early tissue-engineered constructs were exclusively designed using primary cell types that were harvested from tissues that mimicked the functions of the tissues to be repaired or replaced. Tissues can be excised from autologous, allogeneic, or xenogeneic sources. Once explanted, the tissues are enzymatically digested, thus allowing breakdown of the extracellular matrix that essentially holds the tissues together. Cells are also connected to one another via extra-membranous protein ligands which are also digested during this process. Single-cell suspensions are typically filtered or placed in density gradients to remove extraneous cellular and proteinous debris. Cells are then placed in 2-D cultures on tissue culture–grade vessels or maintained in suspension cultures in bioreactors. As described previously in Section III, cells and tissues sourced from either allogeneic or xenogeneic sources are innately immunogenic, and care must be taken when using these cells for transplantation. When transplanted, even in tissue-engineered constructs, these sources present the possibilities of rejection by the graft/implant recipient. Efforts have been made to block recognition of the cells from recipients by encapsulating cells in semipermeable polymer capsules (15). Encapsulation systems can potentially protect transplanted allogeneic and xenogeneic cells from recipient immune responses; however, questions remain as to whether small host immune molecules such as complement proteins and cytokines are capable of passing into the capsules and eliciting rejection responses over time (personal observation).

Alternatives to primary cell types are immortalized cell lines and stem cells. Immortalized cell lines are grown initially from primary cell types that either spontaneously mutate or have been modified in the laboratory using viral transfection in order to render them replication competent. Established cell lines, by definition, are capable of proliferating indefinitely and thus present an inexhaustible supply of cells for tissue engineering. Over time, however, cell lines are prone to spontaneous mutations and therefore recommendations are made to employ these cells during early stages of proliferation. As with primary cells, allogeneic and xenogeneic cell lines pose threats of rejection by the recipient's immune systems (personal observation). Encapsulation of transplanted cells in polymer capsules is a viable option. Nutrients are able to diffuse into the capsules through pores in the semipermeable polymer capsule membranes and cellular products are able to

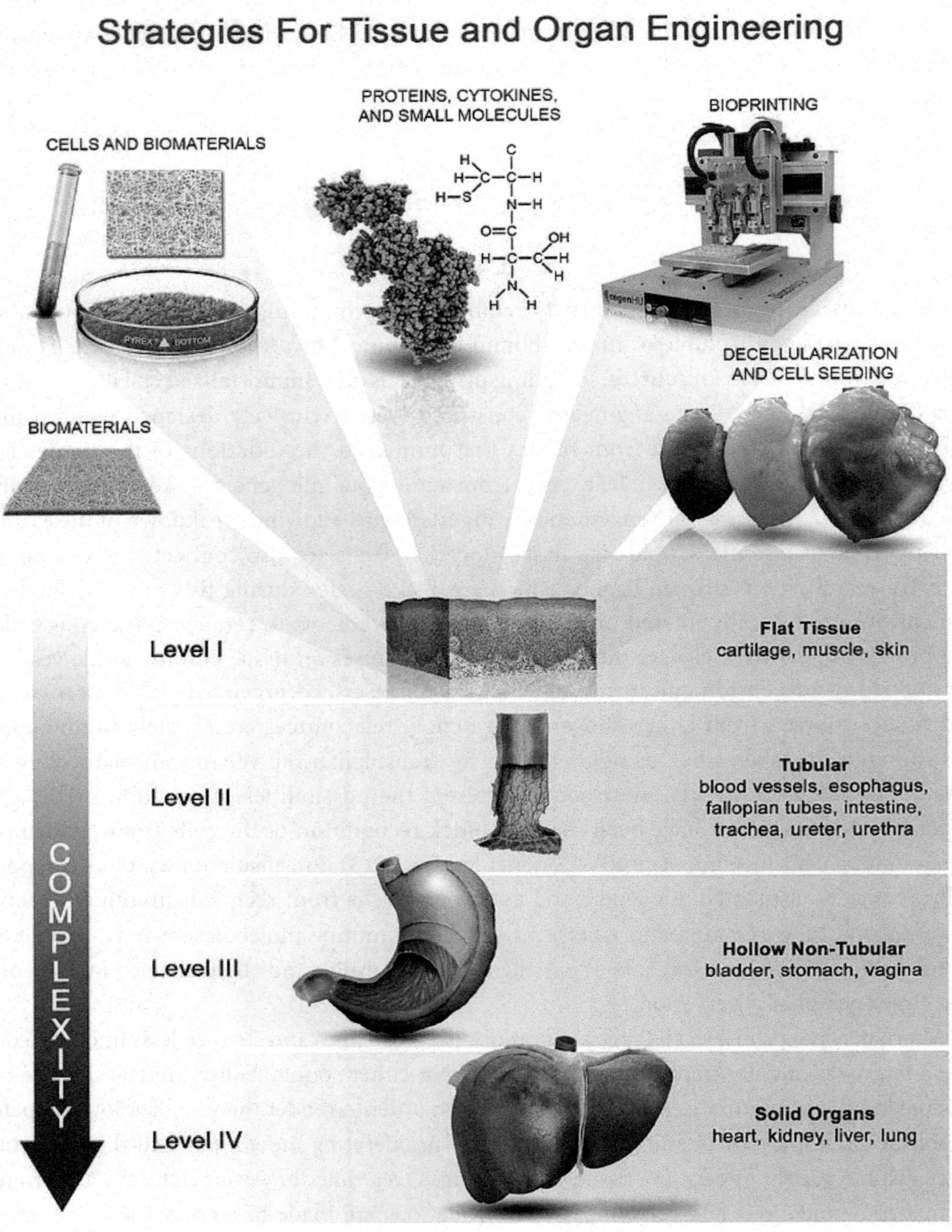

FIGURE 6.2: Strategies for Tissue and Stem Cell Engineering

TABLE 6.1: Cell Types Used for Tissue and Stem Cell Engineering

	Autologous	Allogeneic	Xenogeneic
Primary	Cell donor is also the recipient of the tissue-engineered construct Cells harvested by tissue biopsy followed by enzymatic processing and maintenance in vitro	Cell donors and recipients of tissue-engineered constructs are different individuals within the same species Cells harvested by tissue biopsy followed by enzymatic processing and maintenance in vitro	Cell donors and recipients of tissue-engineered constructs are different species Cells harvested by tissue biopsy followed by enzymatic processing and maintenance in vitro
Cell Line	X	Cells from donors of the same species are transformed into rapidly dividing cell type that can be maintained in vitro for extended period of time	Cells from donors of a different species are transformed into rapidly dividing cell type that can be maintained in vitro for extended period of time
Stem Cells	Cell donor is also the recipient of the tissue-engineered construct Adult stem cell niche is harvested by tissue biopsy followed by enzymatic processing and maintenance in vitro Mature cell types can also be reprogrammed into induced pluripotent stem cells and maintained in vitro	Cell donors and recipients of tissue-engineered constructs are different individuals within the same species Adult stem cell niche is harvested by tissue biopsy followed by enzymatic processing and maintenance in vitro Mature cell types can also be reprogrammed into induced pluripotent stem cells and maintained in vitro Embryonic stem cells can be harvested from developing embryonic blastocyst	Cell donors and recipients of tissue-engineered constructs are different species Adult stem cell niche is harvested by tissue biopsy followed by enzymatic processing and maintenance in vitro Mature cell types can also be reprogrammed into induced pluripotent stem cells and maintained in vitro Embryonic stem cells can be harvested from developing embryonic blastocyst

diffuse out of the pores as well. Caution must be used when encapsulating cell lines, as these cells are capable of continued cell division within the polymer capsules, leading to rupture of capsule membranes and escape of donor cells in the recipient.

Stem cells have become invaluable assets for tissue engineering and regeneration. Discoveries of embryonic as well as adult stem cell niches and induced pluripotent stem cells have significantly increased opportunities for use in science and medicine. All three stem cell types can be obtained from autologous, allogeneic, and xenogeneic sources. As stated previously, embryonic stem cells harvested from blastocysts of developing embryos are pluripotent and carry the highest potential for complete tissue repair and regeneration. Blastocysts form through in situ fertilization of eggs by sperm or in vitro fertilization where sperm is exogenously introduced to eggs in cell culture (16). Once fertilization has occurred, eggs divide into two cell zygotes. Continued cell divisions eventually lead to formation of eight and then sixteen cell blastocysts. In situ formation of blastocysts requires harvesting and further culture in vitro. Each cell of the blastocyst is considered a stem cell and has the potential to divide indefinitely on its own if provided with the appropriate scaffolds and biomolecules. Once cultured in vitro, stem cells proliferate into colonies of genetically identical cells. Theoretically, stem cells have the potential for infinite proliferation in vitro, although spontaneous mutations and differentiation can occur. Terminal differentiation into specific unipotent lineages is achieved by placement of cells onto extracellular matrices and exposure to specialized cocktails of lineage-specific biomolecules (Figure 6.3). Investigators have shown that, prior to terminal differentiation, in vitro cultured stem cells can be grown on 3-D scaffolds and modeled into replacement tissues for later implantation or directly injected back into donors in order to elicit in situ tissue regeneration (Figure 6.4) (17, 18).

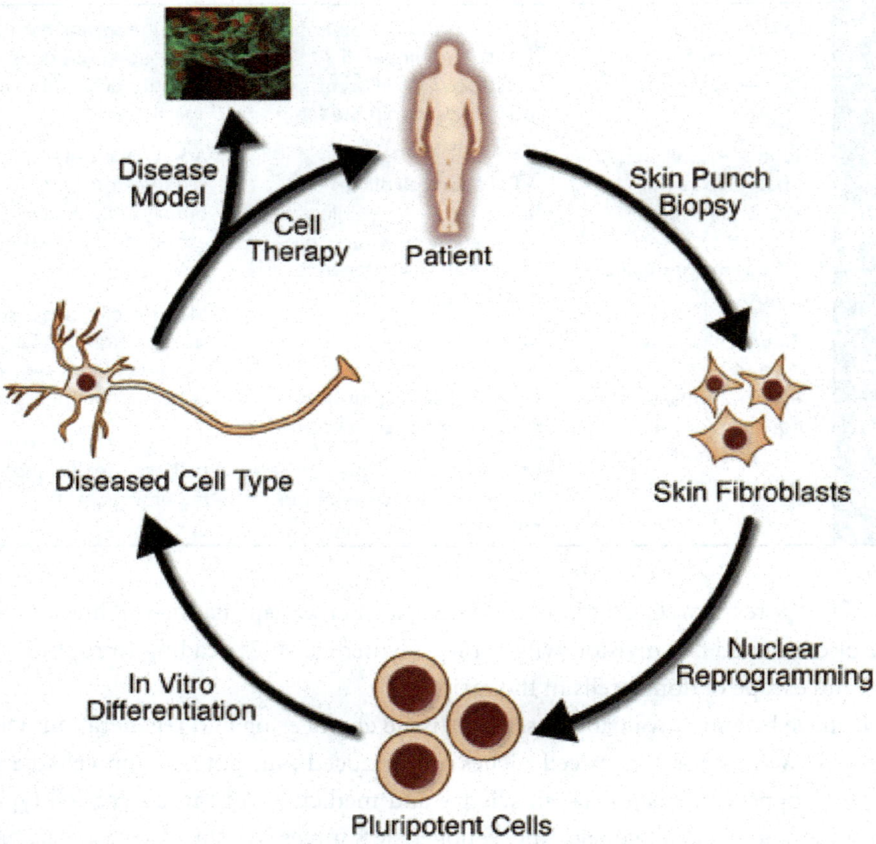

FIGURE 6.3: Autologous Stem Cells for Tissue Engineering

Copyright © 2008 by K.T. Rodolfa, (CC BY 3.0) at https://commons.wikimedia.org/wiki/File:The_steps_of_regenerative_medicine..jpg.

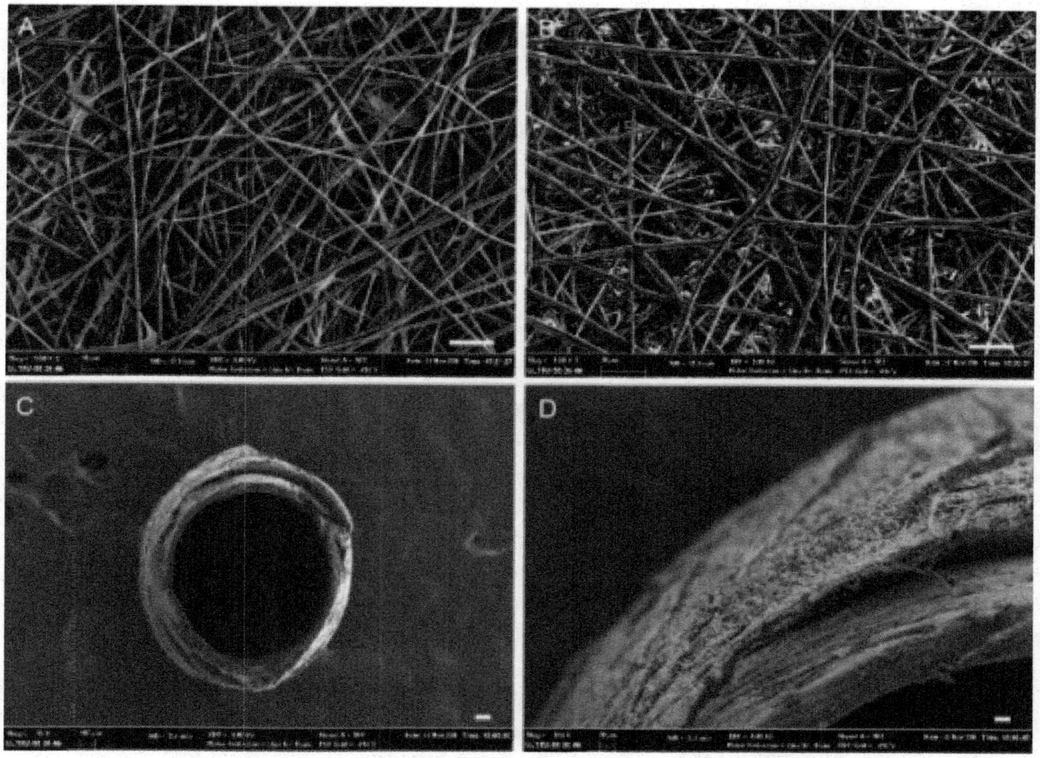

FIGURE 6.4: Electrospun Collagen Fiber Meshes and Conduits
National Institutes of Health / Copyright in the Public Domain.

Due to ethical concerns surrounding the use of embryonic stem cells, alternative cell sources have been sought. Adult stem-like cells are obtained by biopsy from various tissues in the body, including muscle and bone marrow (19). This procedure is immediately followed by enzymatic digestion of tissue into cell suspensions and culture of the suspensions in vitro. Cell cultures are treated in a similar fashion as described for embryonic stem cells, although choices of extracellular matrices and biomolecules vary. One disadvantage of using adult stem-like cells is their intrinsic lack of pluripotency and their inability to differentiate into all cell types in the body. Studies have shown, however, limited capacities of bone marrow–derived stem cells to differentiate into hepatic and cardiac cells both in vitro and in vivo, respectively (20, 21). In fact, clinical trials are being conducted in patients who suffer cardiovascular disease and are being treated with image-guided autologous bone marrow cell injections into affected areas in the heart (14).

The most current and plausible source of stem cells are induced pluripotent stem cells (iPSC) (Figure 6.5) (11). Takahashi and Yamanaka reprogrammed adult murine skin cells by exogenously exposing isolated cells to critical transgenes that forced expressions of four transcription factors (*Oct4*, *Sox2*, *Klf4*, *c-Myc*) that ultimately caused the conversion of these cells into iPSCs (11). Once

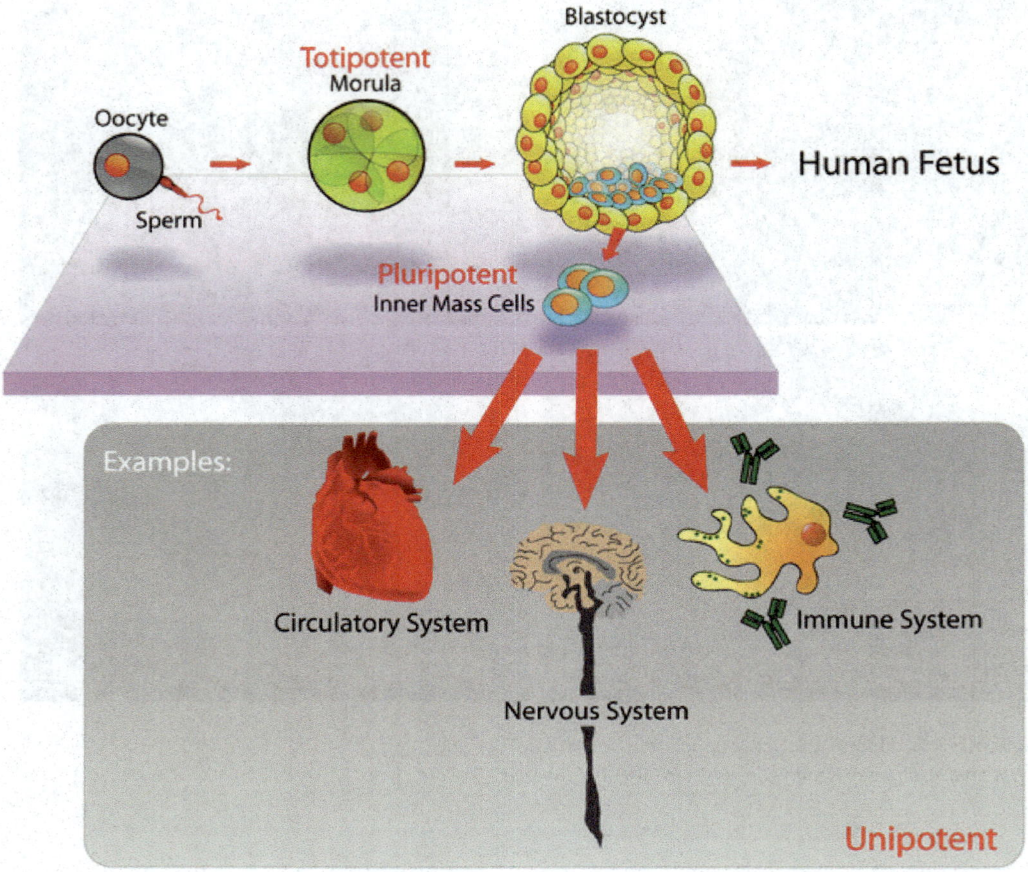

FIGURE 6.5: Stem Cells Harvested from Blastocyst
Copyright © 2006 by Mike Jones, (CC BY-SA 2.5) at https://commons.wikimedia.org/wiki/File:Stem_cells_diagram.png.

iPSCs are in culture, they are exposed to both scaffolds and biomolecules following protocols that are much like those employed for embryonic and adult stem cells. The process of terminal differentiation of iPSCs is called transdifferentiation, based upon the principle that these once differentiated cells are being manipulated to become differentiated cells of a different tissue type. By using adult cells as the primary cell source, investigators are able to obtain stem-like cells that are pluripotent and are not associated with the ethical issues that plague the use of embryonic stem cells. Translation of this technique into clinical practice is not without concerns, however. Random genetic integrations of the transgenes into the genome and tumorigenic risks associated with the use of *c-Myc* have limited the usefulness of iPSCs in clinical settings (22). In order to circumvent this problem, investigators have been able to chemically reprogram adult cells into iPSCs without viral transfection of transgenes (23). An additional concern regarding the use of undifferentiated iPSCs includes immunogenicity of these cells in the recipient following transplantation. Zhao et al. have shown that transplantation of undifferentiated iPSCs results in immune reactions in syngeneic mice as compared to transplantation of undifferentiated embryonic stem cells (24). Screening protocols that determine immunogenicity and tumorigenicity of iPSCs, as well as protocols to enable direct transdifferentiation and avoidance of pluripotency are being investigated as possible solutions (24).

ii. Matrices and Biomolecules

The use of natural materials for repair and replacement of human tissues has been documented for thousands of years. Modern use of natural, synthetic, and combination materials in medicine began in the 1960s; development of specialized biomaterials for tissue engineering purposes began in earnest in the 1980s. Supportive matrices that replace the 3-D structures of native scaffolds are critical for cellular viability in tissue-engineered constructs. In native form, cells—with the exception of blood cells—require complex, interactive matrices in order to survive and function as genotypically and phenotypically intended. Standard cell culture protocols require either 2-D substrates or aggregation of cell clusters in suspension cultures in order to maintain cellular viability. When used to generate functional tissues, cells are removed from these maintenance cultures and grown on 3-D scaffolds in order to replicate the native cellular environment. Scaffolds can either be fabricated from natural polymers such as collagen and decellularized matrices or synthetic polymers such as non-resorbable polyethylene and resorbable poly-l-lactic acid. Collagen is a natural, fibrous protein polymer that is genetically conserved across species. Protein cross-linking with agents such as glutaraldehyde is standard, however, in order to prevent risk of potential antigenicity, as well as to increase mechanical strength of the collagen. Polymer elasticity may be reduced during this process. Collagen-based scaffolds can be fabricated using several techniques, including extrusion and electrospinning. Choice of technique is dependent upon the desired size of polymer fibrils (Figure 6.6) (26). Upon implantation of collagen-cell constructs, cells deposit their own extracellular matrix proteins onto collagen scaffolds, with the ultimate goal being enzymatic degradation of the implanted scaffold and remodeling by transplanted cells. Incorporation by host tissues and vascularization are critical for the success of tissue-engineered implants in terms of cellular viability and functionality. Creation of scaffolds made from natural materials can also be achieved by decellularization of existing tissue structures followed by seeding with desired cell types. Decellularization is a method that employs detergents in order to remove existing cells and significant amounts of antigenic proteins from native tissue and organs (27). Scaffold processing using this method maintains the native structures of entire tissues and organs, thus providing biomimetic structural scaffolds for cell seeding and transplantation. Mechanical integrity can be compromised, however, due to exposure to the cocktail of detergents used.

Synthetic biomaterial scaffolds are widely used in tissue engineering, with resorbable poly-l-lactic acid (PLLA) being the most popular polymer. PLLA can be fabricated into spun polymer fibers, as well as micro- and nano-capsules. Depending upon the molecular weight of the PLLA used and the incorporation of additional resorbable polymers such as poly-lactic-glycolic acid (PLGA), resorption of the scaffold in vivo can be designed to occur over specified time periods. During resorption, water infiltrates the polymer matrix, hydrolytically cleaving polymer bonds, resulting in the formation of water and carbon dioxide. During this time, cells that are seeded on the scaffold deposit their protein matrix and remodel the implant (28). Ultimately, implants become self-supporting and incorporate with adjacent tissues. The first fully implantable bioartificial organ, designed in the United States, was created by Anthony Atala and his laboratory at Wake Forest University in 2001 (29). Nine patients exhibiting varying degrees of urinary incontinence as a result of spina bifida were enrolled in the study. Study endpoints included biological compatibility of implanted materials with patient tissues and implant function as determined by increased compliance and capacitance following implantation. Autologous cells were biopsied from patient's bladders, cultured in vitro, and seeded on PLLA scaffolds. Cell-seeded scaffolds were allowed to

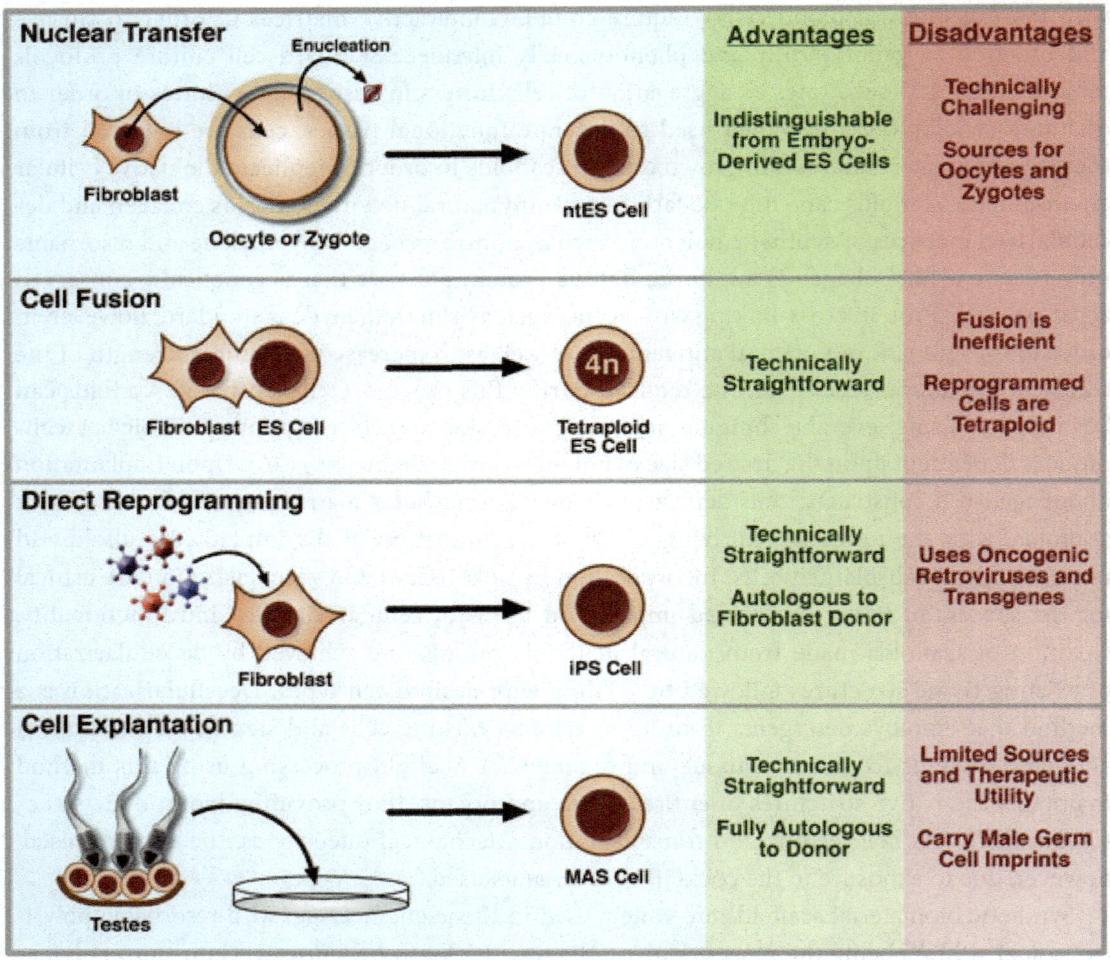

FIGURE 6.6: Methods for Reprogramming Cells for Tissue Engineering

stabilize in vitro and were then anastomosed to the patient's partially resected bladders. In order to establish blood supplies for the cell-seeded implants, scaffolds were wrapped in mesenteric membranes prior to implantation. All patients experienced degrees of success as measured by increases in both bladder compliances and capacities. Vascularization of transplanted cells was provided for by the co-implantation of highly vascularized mesenteric membranes. Adequate oxygenation, nutrition, and metabolite removal are critical for success of all implanted cell-based devices. In addition to providing supplementation by existing vasculature, implant designs are incorporated with soluble and bound biomolecules that support cellular viability and function.

Introduction of tissue-engineered devices into physiological environments elicits wound-healing responses by host tissues. Much like wound healing following surgical incision, several highly orchestrated cellular and biochemical cascades occur that ensure resolution of injury. In order to sustain viabilities and functionalities of implanted tissue-engineered constructs, biochemical signals must be received from soluble and/or bound growth factors. These biochemical signals are essential and are in addition to the structural, chemical, and mechanical signals already being provided for by supportive scaffolds. Although wound-healing responses typically induce hemostasis, acute inflammation, granulation and neovascularization and ultimately tissue remodeling, the presence of "foreign" implants skew these natural immune cascades. Avoidance of chronic inflammatory reactions in relation to tissue-engineered implants is a key priority; vascularization, in direct proximity to the implant, is often difficult to achieve but essential in order to direct appropriate tissue and immune responses. Traditional approaches to growth factor delivery have included injections of soluble molecules and co-implantation of tissue constructs with controlled release systems. Growth factor stability in vivo, high costs, and sequential delivery, however, have limited investigators in optimizing the potentials of tissue-engineered implants. Recently, delivery of growth factors using gene therapy has gained interest, and several studies have been published that cite the advantages of using viral-based gene therapy to deliver bone morphogenetic protein-2 (BMP-2) and insulin growth factor (30). In these studies, human mesenchymal stem cells (MSC) were co-encapsulated in alginate, fibrin, and agarose hydrogels, along with adenoviral vector carrying the cDNA for both growth factors. Hydrogels were implanted in vivo and transfection of MSCs by adenoviral vectors occurred in situ, thus avoiding multiple step processes in vitro (30). Additional delivery methods are under investigation and involve the use of peptide sequences incorporated into biomimetic matrices as single units (31). Guidance of tissue and immune reactions post-implantation is vital for long-term sustainability of implanted tissue-engineered repair and restorative materials. Knowledge of the specificity and sensitivity of tissue microenvironments will ensure intelligent tissue-engineered designs.

Technology on the Edge: 3-D Tissue Structures

In 2002, Professor Makoto Nakamura observed that droplets of ink in a standard inkjet printer were similar in size to human cells (32). Building upon this initial observation, Nakamura replaced the ink in an inkjet printer with mammalian cells and proceeded to adapt the inkjet printer for the purpose of printing human tissues. By 2008, he and his colleagues had developed the first working bioprinter that was capable of printing tubular structures from cellular building blocks (32, 33). Alternate layers of extracellular matrix and mammalian cells were printed drop by drop onto glass plates until a 3-D structure was obtained. Professor Gabor Forgacs proceeded a step further in 2008 and printed functional blood vessels and cardiac tissues from cells harvested from chickens (34). Alternating layers of myocardial and endothelial cells, in

FIGURE 6.7: 3-D Bioprinter
Andrey Ilyin / Copyright in the Public Domain.

addition to layers of collagen-termed "biopaper," were printed onto glass substrates using similar bioprinting methods as Nakamura. Organovo, Inc., a company founded by Professor Nakamura in 2008, successfully created the first blood vessels bioprinted from a single human source. Organovo, Inc. is in the process of preparing bioprinted tissues for clinical trials and plans to provide tissues and organs on demand in the near future.

Another pioneer in the field of bioprinting is Professor Anthony Atala of Wake Forest University. His laboratory has used the technique of bioprinting to create several tissue types including kidney and skin. Furthermore, Atala and his colleagues have established a method to bioprint tissues such as skin in situ (35). In situ deposition of cells along with matrix provides immediate exposure of wounded tissues to reparative cellular and matrix elements and avoids the need for surgical implantation. Repair and replacement of tendons, ligaments, and bone are also being investigated.

Finally, Professor Jeffrey Morgan of Brown University has developed a means to coax mammalian cells into self-assembling into 3-D constructs (36). Morgan and his colleagues have engineered 3-D silicone molds that contain cell culture wells of varying sizes and shapes. Molten agarose is poured into the wells and allowed to solidify. Once removed from the silicone mold, agarose reliefs can be seeded with virtually any cell type, and all cells used thus far have self-assembled into tissue organoids without the use of exogenous extracellular matrix. If removed from the agarose reliefs and combined in culture, the tissue organoids are also capable of fusing into larger constructs. Constructs have the potentials to act as replacement tissues or specific 3-D tissue structures for pharmacological testing.

End-of-Chapter Questions and Answers

1. Stem cells isolated from amniotic fluid are considered:
   Pluripotent
2. Somatic cells that have been genetically engineered to express oncogenic genes that reverse their phenotype/genotype to an undifferentiated state are called:
   Reprogrammed (iPS) stem cells
3. Cells that are considered undifferentiated cells existing within niches of differentiated tissues are:
   Adult stem cells
4. Cell types that express pluripotency and are capable of proliferating in culture without differentiating are:
   Embryonic stem cells
5. Stem cells can initiate immunological rejection reactions if implanted into noncompatible hosts. (T/F)
6. The term used to describe in situ repair or regeneration of damaged tissues is:
   Regenerative medicine

7. The term used to describe in vitro development of the engineered replacement in the laboratory followed by implantation is:
    Tissue engineering
8. Three components necessary for in vitro development of engineered tissues are:
    Scaffold, growth factors, and cells
9. Three species-specific tissue sources that can be harvested for creating in vitro tissue-engineered constructs are:
    Autologous, allogeneic, and xenogeneic
10. Cells harvested for construction of Atala's engineered bladder were isolated from:
    Autologous patient biopsies
11. Parameters that showed improvement following implantation of Atala's tissue-engineered bladder were
    Compliance and capacity
12. The physiological structures critical for normal wound-healing responses, as well as for in vivo survival of engineered tissue constructs are:
    Blood vessels

Works Cited

1. Aras, Onur, and Murat Kazanci. "Production of Collagen Micro- and Nanofibers for Potential Drug-carrier Systems." *Journal of Enzyme Inhibition and Medicinal Chemistry* 30.6 (2015): 1013–016. Web.
2. Chimenti, Isotta, Roberto Gaetani, Lucio Barile, Elvira Forte, Vittoria Ionta, Francesco Angelini, Giacomo Frati, Elisa Messina, and Alessandro Giacomello. "Isolation and Expansion of Adult Cardiac Stem/Progenitor Cells in the Form of Cardiospheres from Human Cardiac Biopsies and Murine Hearts." *Somatic Stem Cells Methods in Molecular Biology* (2012): 327–38. Web.
3. *Clinicaltrials.gov*. US National Institutes of Health, n.d. Web. 16 May 2016.
4. Crubezy, E., P. Murail, L. Girard, and J. P. Bernadou. "False Teeth of the Roman World." *Nature* 391.6662 (1998): 29. Print. 1
5. Cyranoski, David. "Stem Cells Cruise to Clinic." *Nature* 494.7438 (2013): 413. Web.
6. De La Morena, M. T., and R. A. Gatti. *Hematol Oncol Clin North Am*. 25.1 (2011): 1–15. Web.
7. Discher, D. E., D. J. Mooney, and P. W. Zandstra. "Growth Factors, Matrices, and Forces Combine and Control Stem Cells." *Science* 324.5935 (2009): 1673–677. Web.
8. Evans, Martin. "Discovering Pluripotency: 30 Years of Mouse Embryonic Stem Cells." *Nature Reviews Molecular Cell Biology Nat Rev Mol Cell Biol* 12.10 (2011): 680–86. Web.
9. Fell, H. B. "Tissue Culture and Its Contribution to Biology and Medicine." *J. Exp. Biol* 57 (1972): 1–13. Web.

10. Harrison, Rachael H., Jean-Philippe St-Pierre, and Molly M. Stevens. "Tissue Engineering and Regenerative Medicine: A Year in Review." *Tissue Engineering Part B: Reviews* 20.1 (2014): 1–16. Web.
11. Harrison, Ross Granville. "The Outgrowth of the Nerve Fiber as a Mode of Protoplasmic Movement." *J. Exp. Zool. Journal of Experimental Zoology* 9.4 (1910): 787–846. Web.
12. Hou, P., Y. Li, X. Zhang, C. Liu, J. Guan, H. Li, T. Zhao, J. Ye, W. Yang, K. Liu, J. Ge, J. Xu, Q. Zhang, Y. Zhao, and H. Deng. "Pluripotent Stem Cells Induced from Mouse Somatic Cells by Small-Molecule Compounds." *Science* 341.6146 (2013): 651–54. Web.
13. "Human Embryonic Stem Cell Policy Under Former President Bush (Aug. 9, 2001-Mar. 9, 2009)." *Human Embryonic Stem Cell Policy Under Former President Bush (Aug. 9, 2001.* National Institutes of Health, Mar.–Apr. 2009. Web. 16 May 2016.
14. Hussein, S. M. I., K. Nagy, and A. Nagy. "Human Induced Pluripotent Stem Cells: The Past, Present, and Future." *Clin Pharmacol Ther Clinical Pharmacology & Therapeutics* 89.5 (2011): 741–45. Web.
15. Jafari, Maissa, Zahrasadat Paknejad, Maryam Rezai Rad, Saeed Reza Motamedian, Mohammad Jafar Eghbal, Nasser Nadjmi, and Arash Khojasteh. "Polymeric Scaffolds in Tissue Engineering: A Literature Review." *Journal of Biomedical Materials Research Part B: Applied Biomaterials J. Biomed. Mater. Res.* (2015): n. pag. Web.
16. Jung, Jangwook P., Meredith K. Bache-Wiig, Paolo P. Provenzano, and Brenda M. Ogle. "Heterogeneous Differentiation of Human Mesenchymal Stem Cells in 3D Extracellular Matrix Composites." *BioResearch Open Access* 5.1 (2016): 37–48. Web.
17. Kobayashi, M., S. Asakuma, and Y. Fukui. "Blastocyst Production by in Vitro Maturation and Development of Porcine Oocytes in Defined Media Following Intracytoplasmic Sperm Injection." *Zygote* 15.02 (2007): 93. Web.
18. Körbling, Martin, and Zeev Estrov. "Adult Stem Cells for Tissue Repair—A New Therapeutic Concept?" *New England Journal of Medicine N Engl J Med* 349.6 (2003): 570–82. Web.
19. Langer, R., and J. P. Vacanti. "Tissue Engineering." *Science* 260 (1993): 920–26. Web.
20. Macchiarini, Paolo, Philipp Jungebluth, Tetsuhiko Go, M. Adelaide Asnaghi, Louisa E. Rees, Tristan A. Cogan, Amanda Dodson, et al. "Clinical Transplantation of a Tissue-Engineered Airway." *Lancet* 372.9655 (2008): 2023–030. Web.
21. Martino, M. M., P. S. Briquez, A. Ranga, M. P. Lutolf, and J. A. Hubbell. "Heparin-Binding Domain of Fibrin(ogen) Binds Growth Factors and Promotes Tissue Repair When Incorporated within a Synthetic Matrix." *Proceedings of the National Academy of Sciences* 110.12 (2013): 4563–568. Web.
22. Neumann, Alexander J., Josh Schroeder, Mauro Alini, Charles W. Archer, and Martin J. Stoddart. "Enhanced Adenovirus Transduction of HMSCs Using 3D Hydrogel Cell Carriers." *Molecular Biotechnology Mol Biotechnol* 53.2 (2012): 207–16. Web.
23. Obama, Barack. "Executive Order 13505—Removing Barriers to Responsible Scientific Research Involving Human Stem Cells." *The White House.* The White House, 09 Mar. 2009. Web. 16 May 2016.

24. Orive, Gorka, Rosa María Hernández, Alicia R. Gascón, Riccardo Calafiore, Thomas M. S. Chang, Paul De Vos, Gonzalo Hortelano, et al. "Cell Encapsulation: Promise and Progress." *Nature Medicine Nat Med* 9.1 (2003): 104–07. Web.
25. Pearson, Helen. "Scientists Grow Bladder Replacement in Lab." *News@nature* (2006): n. pag. Web.
26. Shamblott, M. J., J. Axelman, S. Wang, E. M. Bugg, J. W. Littlefield, P. J. Donovan, P. D. Blumenthal, G. R. Huggins, and J. D. Gearhart. "Derivation of Pluripotent Stem Cells from Cultured Human Primordial Germ Cells." *Proceedings of the National Academy of Sciences* 95.23 (1998): 13726–3731. Web.
27. Sun, Qisi, Zhonge Zhang, and Zhongjie Sun. "The Potential and Challenges of Using Stem Cells for Cardiovascular Repair and Regeneration." *Genes & Diseases* 1.1 (2014): 113–19. Web.
28. Takahashi, Kazutoshi, and Shinya Yamanaka. "Induction of Pluripotent Stem Cells from Mouse Embryonic and Adult Fibroblast Cultures by Defined Factors." *Cell* 126.4 (2006): 663–76. Web.
29. Wilmut, Ian, Yu Bai, and Jane Taylor. "Somatic Cell Nuclear Transfer: Origins, the Present Position and Future Opportunities." *Phil. Trans. R. Soc. B Philosophical Transactions of the Royal Society B: Biological Sciences* 370.1680 (2015): 2014–0366. Web.
30. Wobma, Holly, and Gordana Vunjak-Novakovic. "Tissue Engineering and Regenerative Medicine 2015: A Year in Review." *Tissue Engineering Part B: Reviews* 22.2 (2016): 101–13. Web.
31. Zhao, Tongbiao, Zhen-Ning Zhang, Zhili Rong, and Yang Xu. "Immunogenicity of Induced Pluripotent Stem Cells." *Nature* 474.7350 (2011): 212–15. Web.
32. Nakamura, M., S. Iwanaga, C. Henmi, K. Arai, and Y. Nishiyama. "Biomatrices and Biomaterials for Future Developments of Bioprinting and Biofabrication." *Biofabrication* 2.1 (2010): 014110. Web.
33. Nakamura, M., S. Iwanaga, C. Henmi, K. Arai, and Y. Nishiyama. "Biomatrices and Biomaterials for Future Developments of Bioprinting and Biofabrication." *Biofabrication* 2. 1 (2010): 014110. Web.
34. Jakab, Karoly, Cyrille Norotte, Francoise Marga, Keith Murphy, Gordana Vunjak-Novakovic, and Gabor Forgacs. "Tissue Engineering by Self-assembly and Bio-printing of Living Cells." *Biofabrication* 2.2 (2010): 022001. Web.
35. Seol, Young-Joon, James J. Yoo, and Anthony Atala. "Bioprinting of Three-Dimensional Tissues and Organ Constructs." *Essentials of 3D Biofabrication and Translation* (2015): 283–92. Web.
36. Svoronos, Alexander A., Nalin Tejavibulya, Jacquelyn Y. Schell, Vivek B. Shenoy, and Jeffrey R. Morgan. "Micro-Mold Design Controls the 3D Morphological Evolution of Self-Assembling Multicellular Microtissues." *Tissue Engineering Part A* 20.7–8 (2014): 1134–144. Web.

CPSIA information can be obtained
at www.ICGtesting.com
Printed in the USA
LVOW06s2149310816
502691LV00004B/29/P

9 781634 876599